GOVERNMENT PROCEDURES AND OPERATIONS

THE FDA AND GENERALLY RECOGNIZED AS SAFE (GRAS) SUBSTANCES

Government Procedures and Operations

Additional books in this series can be found on Nova's website under the Series tab.

Additional E-books in this series can be found on Nova's website under the E-books tab.

Food and Beverage Consumption and Health

Additional books in this series can be found on Nova's website under the Series tab.

Additional E-books in this series can be found on Nova's website under the E-books tab.

GOVERNMENT PROCEDURES AND OPERATIONS

THE FDA AND GENERALLY RECOGNIZED AS SAFE (GRAS) SUBSTANCES

JAMES C. MONAHAN
EDITOR

Nova Science Publishers, Inc.
New York

Additional color graphics may be available in the e-book version of this book.

Library of Congress Cataloging-in-Publication Data

The FDA and Generally Recognized as Safe (GRAS) substances / editors, James C. Monahan.
p. cm.
Includes index.
ISBN 978-1-61209-924-8 (hardcover)
1. United States. Food and Drug Administration. 2. Food adulteration and inspection--Law and legislation--United States. I. Monahan, James C.
KF3878.F33 2011
344.7304'232--dc22

2011004604

Published by Nova Science Publishers, Inc. + New York

CONTENTS

PREFACE

The FDA, which is responsible for ensuring the safety of most of the U.S. food supply, is not required to review substances, such as spices and preservatives, added to food that are generally recognized as safe (GRAS) for their intended use. Currently, companies may determine a substance is GRAS without FDA's approval or knowledge. However, a few substances previously considered GRAS have later been banned and concerns have been raised about the safety of other GRAS substances, including those containing engineered nanomaterials. This new book provides an overview of the FDA's regulation of GRAS substances, which are exempt from the premarket approval process for food additives, and discusses the FDA's proposed rule.

Chapter 1- The Food and Drug Administration (FDA), which is responsible for ensuring the safety of most of the U.S. food supply, is not required to review substances, such as spices and preservatives, added to food that are generally recognized as safe (GRAS) for their intended use. Currently, companies may determine a substance is GRAS without FDA's approval or knowledge. However, a few substances previously considered GRAS have later been banned; and concerns have been raised about the safety of other GRAS substances, including those containing engineered nanomaterials, materials manufactured at a tiny scale to take advantage of novel properties. GAO was asked to review the extent to which (1) FDA's oversight of new GRAS determinations helps ensure the safety of these substances, (2) FDA ensures the continued safety of current GRAS substances, and (3) FDA's approach to regulating engineered nanomaterials in GRAS substances helps ensure the safety of the food supply. GAO reviewed FDA data on GRAS substances and interviewed a range of stakeholders, among other things.

Chapter 2- The use of carbon monoxide (CO) in the packaging of meat and fish has generated considerable debate. The presence of CO results in the meat turning a bright red color that lasts longer than the color in untreated meat. Additionally, fish treated with CO gain a fresher appearance and a red tint. The meat industry, consumer groups, scientists, and policy makers disagree as to whether the use of CO in meat and fish packaging should be regulated by the Food and Drug Administration (FDA) and the United States Department of Agriculture (USDA), through labeling or otherwise, and whether CO should be a substance Generally Recognized As Safe (GRAS) under current and proposed FDA rules.

Chapter 3- This guidance represents the Food and Drug Administration's (FDA's) current thinking on this topic. It does not create or confer any rights for or on any person and does not operate to bind FDA or the public. An alternative approach can be used if such approach satisfies the requirements of the applicable statutes and regulations.

In: The FDA and Generally Recognized as Safe … ISBN: 978-1-61209-924-8
Editor: James C. Monahan, pp. 1 - 68

Chapter 1

FDA SHOULD STRENGTHEN ITS OVERSIGHT OF FOOD INGREDIENTS DETERMINED TO BE GENERALLY RECOGNIZED AS SAFE (GRAS)*

United States Government Accountability Office

WHY GAO DID THIS STUDY

The Food and Drug Administration (FDA), which is responsible for ensuring the safety of most of the U.S. food supply, is not required to review substances, such as spices and preservatives, added to food that are generally recognized as safe (GRAS) for their intended use. Currently, companies may determine a substance is GRAS without FDA's approval or knowledge. However, a few substances previously considered GRAS have later been banned; and concerns have been raised about the safety of other GRAS substances, including those containing engineered nanomaterials, materials manufactured at a tiny scale to take advantage of novel properties. GAO was asked to review the extent to which (1) FDA's oversight of new GRAS determinations helps ensure the safety of these substances, (2) FDA ensures the continued safety of current GRAS substances, and (3) FDA's approach to regulating engineered nanomaterials in GRAS substances helps ensure the safety of the food supply. GAO reviewed FDA

* This is an edited, reformatted and augmented edition of a United States Government Accountability Office publication, Report GAO-10-246, dated February 2010.

data on GRAS substances and interviewed a range of stakeholders, among other things.

What GAO Recommends

GAO recommends that FDA take steps to better ensure the safety of GRAS substances, including developing a strategy to require any company that conducts a GRAS determination to provide FDA with basic information about it. FDA generally agreed, while raising concerns about certain aspects of several of the recommendations.

What GAO Found

FDA's oversight process does not help ensure the safety of all new GRAS determinations. FDA only reviews those GRAS determinations that companies submit to the agency's voluntary notification program—the agency generally does not have information about other GRAS determinations companies have made because companies are not required to inform FDA of them. Furthermore, FDA has not taken certain steps that could help ensure the safety of GRAS determinations, particularly those about which the agency has not been notified. FDA has not issued guidance to companies on how to document their GRAS determinations or monitored companies to ensure that they have conducted GRAS determinations appropriately. Lastly, FDA has yet to issue a final regulation for its 1997 proposed rule that sets forth the framework and criteria for the voluntary notification program, potentially detracting from the program's credibility.

FDA is not systematically ensuring the continued safety of current GRAS substances. While, according to FDA regulations, the GRAS status of a substance must be reconsidered as new scientific information emerges, the agency has not systematically reconsidered GRAS substances since the 1980s. FDA officials said they keep up with new developments in the scientific literature and, on a case-by-case basis, information brought to the agency's attention could prompt them to reconsider the safety of a GRAS substance. However, FDA has largely not responded to concerns about GRAS substances, such as salt and the trans fats in partially hydrogenated vegetable oils, that individuals and consumer groups have raised through 11 citizen petitions submitted to the agency between 2004 and 2008. In fact, FDA has decided on the validity of these concerns in only

1 of 11 cases. In addition, FDA does not know to what extent, or even whether, companies track evolving scientific information about their GRAS substances.

FDA's approach to regulating nanotechnology allows engineered nanomaterials to enter the food supply as GRAS substances without FDA's knowledge. While some uses of engineered nanomaterials have the potential to help ensure food safety, uncertainties remain about how to determine their safety in food. After reviewing the uncertainties associated with the safety of engineered nanomaterials, FDA has decided that it does not need additional authority to regulate products containing such materials. Rather, FDA encourages, but does not require, companies considering using engineered nanomaterials in food to consult with the agency regarding whether such substances might be GRAS. Because GRAS notification is voluntary and companies are not required to identify nanomaterials in their GRAS substances, FDA has no way of knowing the full extent to which engineered nanomaterials have entered the U.S. food supply as part of GRAS substances. In contrast to FDA's approach, all food ingredients that incorporate engineered nanomaterials must be submitted to regulators in Canada and the European Union before they can be marketed.

Abbreviations

EAFUS	Everything Added to Food in the United States
EPA	Environmental Protection Agency
FDA	Food and Drug Administration
GRAS	generally recognized as safe

February 3, 2010

The Honorable Tom Harkin
Chairman
Committee on Health, Education, Labor and Pensions
United States Senate

The Honorable Rosa L. DeLauro
Chair
Subcommittee on Agriculture, Rural Development, Food and Drug Administration and Related Agencies
Committee on Appropriations
House of Representatives

The Food and Drug Administration (FDA), which is responsible for ensuring the safety of most of the U.S. food supply, does not review many of the substances added to food that manufacturers determine to be generally recognized as safe (GRAS) under the conditions of their intended use. Manufacturers add these substances—hundreds of spices and artificial flavors, emulsifiers and binders, vitamins and minerals, and preservatives—to enhance a food's taste, texture, nutritional content, or shelf life. GRAS substances can be marketed without FDA's approval or even its knowledge because such substances are generally recognized among qualified experts as having been shown, through scientific procedures or experience based on common use, to be safe. Some consider GRAS substances to warrant less oversight because they generally pose a relatively low level of threat to public health. However, a few substances previously assumed to be GRAS, such as cyclamate salts, have later been banned; and more recently, consumer groups have raised concerns about the safety of certain other GRAS substances, such as salt and trans fats in partially hydrogenated vegetable oils.

The Federal Food, Drug, and Cosmetic Act exempts GRAS substances from the act's general requirement that companies obtain FDA approval before marketing food additives (substances, when used as intended, reasonably expected to become a component or otherwise affect the characteristics of food). This exemption allows companies, without the approval of FDA, to determine whether there is enough support to claim a substance is GRAS.[1] For a company to claim a substance is GRAS, it must conclude that there is common knowledge about the safety of the substance among experts qualified by scientific training and experience to evaluate its safety. Under a program set forth in a rule FDA proposed in 1997, companies may voluntarily submit information on a substance they conclude is GRAS to the agency's GRAS notification program for review. After reviewing this information, FDA may state in a letter that it has no questions about the company's GRAS determination. However, once a company—domestic or foreign—concludes that a substance is GRAS, it may market that substance as GRAS without informing FDA. Anyone may request that FDA change or create an agency regulation through a citizen petition, and groups and individuals submitted almost 50 such petitions to FDA related to GRAS substances from 1975 through 2008.

In recent years, concerns have been raised about the potential for engineered nanomaterials to be considered GRAS for use in food and food packaging until more is known about their risks. Engineered, or manufactured, nanomaterials are created through nanotechnology—the creation and manipulation of materials at a very small (molecular) scale that enhances certain of the resulting nanomaterials' physical properties. Applications of these nanomaterials in food and food packaging have the potential to benefit food safety. For example, antimicrobial nanofilms—thin layers of substances that hamper the growth of bacteria and fungi—in food packaging could decrease foodborne pathogens. While the underlying chemical structure of a substance is not changed by the engineering process, its physical properties may change. For example, while the chemical structure of salt is the same—whether at its natural scale or at the nanoscale—it may be possible to reduce the amount of salt in a product by using it at the nanoscale to coat other particles.

In this context, you asked us to review FDA's oversight of GRAS substances. This report examines the extent to which (1) FDA's oversight of new GRAS determinations helps ensure the safety of these substances, (2) FDA ensures the continued safety of current GRAS substances as new scientific information emerges, and (3) FDA's approach to regulating engineered nanomaterials in GRAS substances helps ensure the safety of the food supply. We also provide additional information on the safety of two GRAS substances—salt and trans fats in partially hydrogenated vegetable oils—including the views of the *Dietary Guidelines for Americans*,[2] *Dietary Reference Intakes*,[3] and FDA's Food Advisory Committee; this information is discussed in appendix I.

To review FDA's oversight of GRAS determinations, we compiled and analyzed data on FDA's voluntary notification program from 1998, the first year a GRAS notice was submitted, through 2008. Specifically, we used FDA's GRAS Notice Inventory database. To assess the reliability of the data used in this report from this source, we reviewed related documentation, examined the data to identify obvious errors or inconsistencies, and worked with agency officials to identify any data problems and steps they took to ensure the reliability of the data. Based on this examination, we concluded that these data were sufficiently reliable for the purposes of this report. We also reviewed laws and regulations regarding GRAS substances. To review the extent to which FDA ensures the continued safety of GRAS substances, we examined the 11 citizen petitions related to GRAS substances that were submitted to FDA during the recent 5-year period from 2004 through 2008 and gathered information from FDA officials regarding the agency's response to these petitions. Finally, to review FDA's approach to regulating engineered nanomaterials as GRAS, we evaluated the agency's policies and guidance to companies, and collected information about the activities of foreign governments—namely, Canada and the European Union—that have been particularly active in considering regulation of engineered nanomaterials in food. In addition, to address all of our objectives, we interviewed a wide range of stakeholders, including officials from FDA, industry and trade organizations, consumer advocacy groups, academia, and foreign governments. Appendix II provides a more detailed description of our scope and methodology.

We conducted this performance audit from October 2008 to February 2010, in accordance with generally accepted government auditing standards. Those standards require that we plan and perform the audit to obtain sufficient, appropriate evidence to provide a reasonable basis for our findings and conclusions based on our audit objectives. We believe that the evidence obtained provides a reasonable basis for our findings and conclusions based on our audit objectives.

BACKGROUND

In 1958, Congress amended the Federal Food, Drug, and Cosmetic Act to state that food bearing or containing any unsafe food additive shall be deemed adulterated and that a food additive is considered unsafe unless, among other things, it and its use conform to a regulation prescribing the conditions of its safe use. A regulation prescribing the conditions for safe use could be obtained through a process established in the 1958 amendment. That process describes the data for companies to include in a petition for a proposed regulation (known as a food additive petition), a time frame and standards for agency review, and an opportunity for public comment. Under this process, as further prescribed in FDA regulations, the safety of an additive does not need to be established with absolute certainty; instead, the regulations provide a science-based standard of safety, requiring a reasonable certainty that no harm will result from the intended use of an additive. If FDA finds a food additive to be safe for its intended uses, the agency will issue a regulation specifying those uses.

The 1958 amendment defines "food additive" as any substance the intended use of which results, or may reasonably be expected to result, directly or indirectly, in its becoming a component of or otherwise affecting the character of any food. However, this definition excludes substances that are generally recognized as safe under the conditions of their intended use, as shown through scientific procedures or based on common usage in food prior to 1958. Therefore, unlike food additives, a GRAS substance is not considered unsafe in the absence of a regulation prescribing its safe use, allowing a company to market food containing the substance without FDA approval.

Since the 1958 amendment, FDA has taken a variety of actions to determine the GRAS status of substances used in food, as figure 1 illustrates.

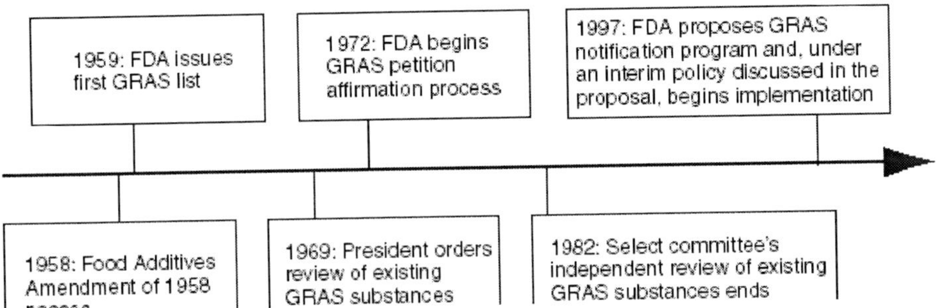

Source: GAO analysis of FDA information.

Figure 1. Timeline of Significant Events in FDA's GRAS Program.

Shortly after the passage of the 1958 amendment, FDA clarified the regulatory status of many food substances that were used in food prior to 1958 and amended its regulations to include a list of food substances, commonly referred to as the GRAS list, that, when used for the purposes indicated and in accordance with current good manufacturing practices, are GRAS. However, FDA also acknowledged the extreme difficulty in listing all substances that are GRAS for their intended use, stating in its regulations that such an effort would be "impracticable." FDA added other categories of substances—for example, spices, seasonings, and flavorings—to the GRAS list in subsequent rule makings.

During the late 1960s, new scientific information raised questions about the safety of cyclamate salts, a class of sweeteners that FDA previously considered GRAS. As a result, FDA contracted for an independent review, by contemporary standards, of the available safety information related to substances it considered GRAS. If the review confirmed that the use of a particular substance was GRAS, FDA would issue a new regulation, affirming that finding. At about the same time as this review began, FDA also established procedures—referred to as the petition affirmation process—whereby companies could petition FDA to affirm the GRAS status of substances not covered as part of the review. To the extent that companies voluntarily submitted petitions to FDA as part of the affirmation process, the agency became aware of companies' independent GRAS determinations.

Under FDA's GRAS petition affirmation process—as described in FDA's 1997 proposal for a new GRAS program—the agency (1) published a notice in the *Federal Register*; (2) requested comments on the GRAS petition; (3) comprehensively reviewed the safety of the substance; (4) drafted a detailed explanation of why the use is GRAS; and (5) published that explanation in the

Federal Register. Each petition resulted in a rule making that allowed the public to comment on the proposed rule and FDA to respond to these comments before issuing the final rule.

In 1997, citing the resource intensive process required to conduct the petition affirmation process, FDA proposed a new GRAS program that, among other things, would eliminate the rule-making steps under the affirmation process. FDA's proposed rule would allow companies that had made a GRAS determination to apply to FDA for its review under a voluntary notification program. Under this new program, in which FDA invited companies to participate under an interim policy discussed in the proposed rule, FDA no longer affirms the GRAS status of a substance. Rather, once FDA completes its review of a company's notice of a GRAS determination, it informs the company in a letter of one of the following three responses:

- FDA has no questions about the company's conclusion that the substance is GRAS (referred to as a no questions letter);
- FDA concludes that the notice does not provide a sufficient basis for a determination that the substance is GRAS because, for example, the notice does not include appropriate data or the available data raise questions about the safety of the substance; or
- FDA has, at the company's request, ceased to evaluate the GRAS notice.

In proposing the rule, FDA asserted that, from the companies' standpoint, the proposed voluntary notification program was simpler than the GRAS petition affirmation process and, therefore, could provide an incentive for manufacturers to inform FDA of their GRAS determinations. A flowchart presenting steps in the voluntary notification program is found in appendix III.

In the proposal, FDA invited companies to submit notifications to the GRAS notification program described in the proposed rule until it published a final rule. FDA did not formally terminate the petition affirmation process, but has stated it no longer commits resources to the process. Because the rule has not been made final, FDA has operated the interim GRAS notification program under this proposed rule since 1997.

One way for citizens to question GRAS determinations is through citizen petitions. FDA regulations establish procedures for petitioning FDA to issue, amend, or revoke a regulation or order; or take or refrain from taking any other form of administrative action. If a petition appears to meet the requirements for submission, FDA is required to furnish a response within 180 days of its receipt, if not sooner. In reviewing a petition, FDA may use the following procedures: (1)

conferences, meetings, discussions, and correspondence; (2) a hearing; (3) a *Federal Register* notice requesting information and views; (4) a proposal to issue, amend, or revoke a regulation; or (5) other specifically established procedures. The record of any of these steps becomes part of the administrative record for the petition. FDA must generally respond to a petition within 180 days and (1) approve it and, therefore, concurrently take appropriate action implementing the approval (for example, publishing a *Federal Register* notice); (2) deny it; or (3) provide a tentative response, indicating why the agency has been unable to decide on it (for example, because of the existence of other agency priorities, or a need for additional information). FDA's tentative response may indicate its likely ultimate decision and may specify when this final decision is to be provided. FDA must notify the petitioner in writing of its decision. FDA may grant or deny citizen petitions, in whole or in part, and may take other action as the petition warrants.

Researchers are studying nanotechnology—the creation and manipulation of materials at a very small scale—to explore its many potential uses in food manufacturing, including uses potentially beneficial to food safety. Nanotechnology can involve processes, materials, and applications that span physical, chemical, biological, engineering, and electronic sciences. Although definitions of nanotechnology vary, the National Nanotechnology Initiative, a federal program established in 2001 to coordinate nanotechnology research and development, has defined it as the understanding and control of matter between 1 and 100 nanometers, known as the nanoscale.[4] A nanometer is one billionth of a meter, a size that can best be understood by comparison to other very small objects—for example, the diameter of a human hair is approximately 80,000-100,000 nanometers, that of a red blood cell is approximately 8,000 nanometers, and that of a typical virus between 80 and 120 nanometers. Nanomaterials are materials at the nanoscale; or they can be larger, if they retain the characteristics of nanomaterials, such as novel properties compared with the same materials at their natural scale. Nanomaterials can be found in nature, such as in soot; or be an inadvertent product of traditional manufacturing practices, such as nanomaterials present as emulsions in homogenized milk or mayonnaise. Engineered nanomaterials, however, are materials that are deliberately manufactured to take advantage of the novel properties that occur at the nanoscale.

When reduced to the nanoscale, substances can take on novel properties that allow them to be used in applications for many different purposes across different industries. For example, by reducing materials to the nanoscale, the materials' surface area is increased, which can affect the nanomaterials' ability to react with other substances. In addition, nanomaterials may have a greater ability to move across biological membranes. Because of these properties, nanotechnology may offer technological advancements in food packaging and storage that enhance the shelf life of fresh foods. Applications of nanotechnology may also offer more efficient nutrient delivery. Such applications in food and food packaging are relatively new, and FDA has approved only a few such uses in food contact substances so far.

FDA's Oversight Process Does Not Help Ensure the Safety of All New GRAS Determinations

FDA reviews those GRAS determinations that companies choose to submit to the voluntary notification program. However, FDA generally does not have information about other GRAS determinations because companies are not required to inform the agency of their GRAS determinations. Furthermore, FDA has not taken certain steps that could help ensure the safety of GRAS determinations, particularly those for which the agency has not been notified. Notably, a trade association routinely informs FDA of its GRAS determinations, even though it does not submit notices to FDA's voluntary notification program.

FDA Reviews Those GRAS Determinations that Companies Choose to Submit

From 1998—the first year a company submitted a notice of a GRAS determination—through 2008, companies chose to submit 274 GRAS determinations to FDA under the 1997 proposed voluntary notification program, or about 25 annually. According to FDA, it has received notices for substances such as carbohydrates, lipids, proteins, and chemicals. At any given time, FDA may have pending notices—notices under review for which FDA has not yet issued a final opinion. Table 1 shows the status of FDA's responses to these GRAS notices.

FDA encourages companies to meet with agency officials before formally submitting their notices of GRAS determinations. In this presubmission meeting, FDA informally reviews the scientific information the company plans to submit. The company may forgo submitting the notice for several reasons, including if the agency anticipates that the scientific support would not meet the required safety standard for GRAS substances. Within 30 days of receiving a company's notice of a GRAS determination, FDA informs the company in writing of the date on which the notice was received. FDA convenes a group of staff—referred to as its GRAS Notice Review Team—to evaluate the evidence the company submitted with its notice. FDA then evaluates whether the submitted notice provides a sufficient basis for a GRAS determination and whether information in the notice, or otherwise available to FDA, raises questions about whether the substance is GRAS for its intended use. If, during the review, FDA finds that the company's GRAS notice lacks sufficient information, it gives the company the opportunity to provide supplemental information. However, once a company concludes that a substance is GRAS, it may market the substance, even if FDA finds that the notice does not provide a sufficient basis for a GRAS determination.

Table 1. FDA Responses to GRAS Notices Received under Voluntary Notification Program, 1998-2008.

FDA response	Number of response letters	Percentage of responses
FDA has no questions	211	77
Notice does not provide a basis for a GRAS determination	16	6
At company's request, FDA ceased to evaluate the notice	41	15
Pending	6	2
Total	**274**	**100**

Source: GAO analysis of FDA data.

Notes: (1) In 16 cases, companies resubmitted the notice after withdrawing it and in 6 cases, companies resubmitted the notice after FDA stated the notice did not provide a basis for a GRAS determination; FDA assigns a new GRAS notice number when substances are resubmitted. (2) Data are as of October 29, 2009.

Table 2. FDA Performance in Meeting 180-Day Time Frame for Completing Review of GRAS Notices, 1998-2008.

Amount of time for FDA to complete review after receiving notice	Number of notices	Percentage of notices
Within 180 days	175	64
181 or more days	93	34
Pending[a]	6	2
Total	**274**	**100**

Source: GAO analysis of FDA data.

Notes: (1) Until 2001, FDA's goal was to complete its review of companies' notices within 90 days of receipt. (2) Data are as of October 29, 2009.

[a]In these cases, FDA has not completed its review of the company's notice.

In the 1997 proposal, FDA indicated that it planned to complete its review of companies' notices within 90 days of receipt, but stated that it would determine whether its experience in administering such notices suggested modifications to the proposed procedures. In 2001, FDA lengthened this time frame to 180 days for most notices because the agency found that it took longer than anticipated to review the notices, according to FDA officials. As shown in table 2, FDA has met the latter time frame for about 64 percent of notices over the course of the program.

According to FDA officials, delays in meeting its 180-day time frame for review of a GRAS notice can occur for various reasons. For example, in its internal guidance to the GRAS Notice Review Team, FDA stated that the 180-day time frame is contingent on the degree of the notice's complexity—the 180-day time frame is designed as an achievable goal for GRAS notices that are of low to moderate complexity. FDA officials also explained that delays can occur in meeting the time frame because the agency sometimes requests additional information from companies during its review. FDA expects that companies would respond to such requests in a timely manner; but they do not always do so, and FDA does not require a response within a certain amount of time. If a company does not respond to the agency's request for additional information, FDA may eventually contact the company and ask it to consider withdrawing its application. Recently, as shown in figure 2, FDA has met the 180-day time frame less frequently than in the past—excluding pending notices, the agency met this time frame in 44 percent of cases from 2005 through 2008, while it met this time frame in 79 percent of cases from 1998 through 2004. According to FDA officials, this delay in reviews has generally occurred because of budget limitations, increased demands on staff time, and loss of key staff for the office that conducts the reviews. Agency officials said they anticipate that their ability to complete reviews in a timely manner should improve in the future because they have recently hired additional staff, including some staff under contract with limited terms, for that office.

FDA officials and industry representatives explained that a company that manufactures a GRAS substance has incentive to submit a notice to the voluntary notification program for review because FDA's no questions letter improves the company's ability to market its GRAS substance to companies that purchase GRAS substances as ingredients for their food products. Companies that purchase GRAS substances may require or prefer that these substances have been reviewed by FDA. One company's representative explained that FDA's voluntary notification program is also beneficial because FDA scientists' review provides the company with additional assurance of safety.

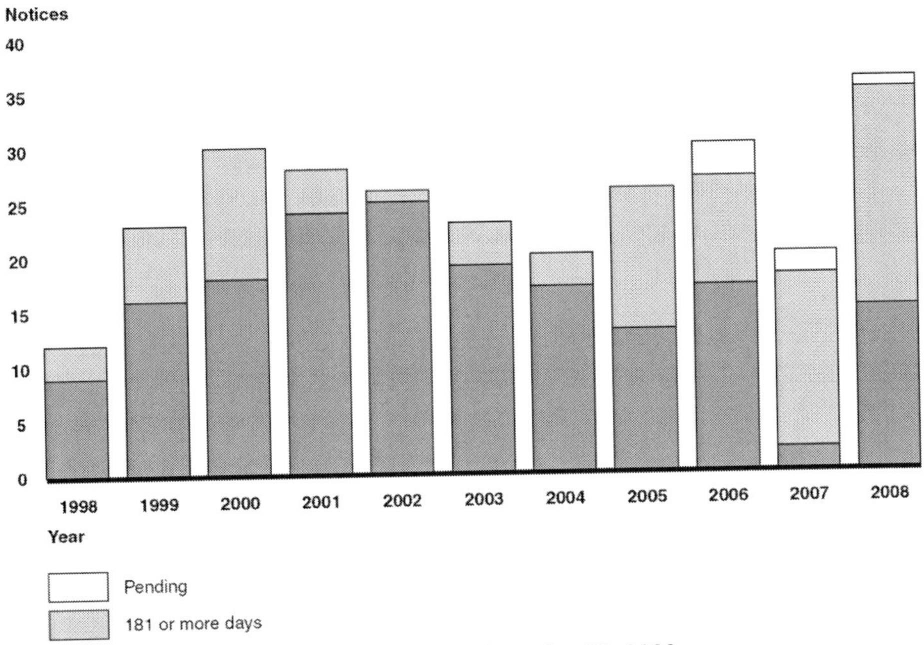

Note: Data on pending reviews are current as of October 29, 2009.

Figure 2. FDA's Annual Performance in Meeting the 180-Day Time Frame for Completing GRAS Notice Reviews, 1998-2008.

FDA has also taken steps to make information about the GRAS notification program available to the public by posting its inventory of all GRAS notices FDA has received on its Web site. The Web site describes FDA's response to each notice as either (1) FDA has no questions; (2) notice does not provide a basis for a GRAS determination; or (3) at the company's request, FDA ceased to evaluate the notice. The Web site also provides a hyperlink to the agency's response letter and, in many cases, to the GRAS notice as well. By placing information about the GRAS notice and its response on its Web site, FDA enhances the ability of Congress, stakeholders, and the general public to be better informed about GRAS substances.

FDA Generally Has No Information about GRAS Determinations That Are Not Submitted to Its Notification Program

Although FDA's voluntary notification program allows the agency to review those GRAS determinations companies submit, FDA generally does not have information about other GRAS substances in the marketplace because companies are not required to provide information to FDA regarding their GRAS determinations. For example, officials representing one international marketer of food indicated the company makes about 5 GRAS determinations each year without notifying FDA. These are usually new uses of substances that have been deemed GRAS for other uses. In another case, a company began marketing a purified version of stevia, a plant-based sweetener, as a GRAS substance before submitting a notice to FDA and before FDA had indicated it had no questions about other GRAS notices related to stevia.[5]

Once a GRAS substance has entered the marketplace, FDA would find it difficult to identify that substance as the potential source of a food safety problem, especially if FDA is unaware that the substance has been determined to be GRAS. Food products may contain numerous ingredients, including GRAS substances, making it difficult, if not impossible, for public health authorities to attribute a food safety problem to a specific GRAS substance. Moreover, while FDA receives reports of adverse reactions to food, it is difficult to clearly identify any specific GRAS substance as the likely cause of a foodborne illness from these reports. Because of the difficulty of identifying GRAS substances as the source of food safety problems after they have entered the food supply, FDA's oversight of their safety would be improved if companies were required to make the agency aware of their GRAS determinations. In this way, FDA would already have at least some information in its databases about GRAS substances, which could help its investigations of food safety problems.

Furthermore, without overseeing all companies' GRAS determinations, FDA is less informed about the nation's food supply and consumers' cumulative dietary exposure to GRAS substances—both of which were viewed by FDA as beneficial potential outcomes of its 1997 proposal. FDA maintains a database named Everything Added to Food in the United States (EAFUS). Despite its name, FDA acknowledges that this database is incomplete because companies are not required to participate in the GRAS notification program or even inform FDA of their GRAS determinations, and FDA officials cannot estimate the number of determinations that occur about which they are not notified. Although approximately 180 companies submitted notices of GRAS determinations to FDA's voluntary notification program from 1998 through 2008, the agency does not know to what extent these or other companies made GRAS determinations during this period but chose not to notify the agency. Without information about all GRAS determinations, FDA has less awareness of substances in the nation's food supply and less knowledge of the potential cumulative dietary exposure of GRAS substances. However, FDA officials said that EAFUS incorporates information on most food ingredients, and they indicated they are not significantly concerned about missing GRAS substances in the database because, as some food scientists have indicated, GRAS substances generally pose a relatively low risk to public health.

The safety of imported food products, including those containing GRAS substances, is also a matter of concern. GRAS substances may be manufactured anywhere in the world and FDA does not track where they are manufactured. FDA has stated that it knows of no other country that has a law comparable to the GRAS provision of the Federal Food, Drug, and Cosmetic Act. While other countries do not have this GRAS provision, GRAS substances brought into the United States can be manufactured anywhere if in compliance with U.S. food safety law and FDA regulations. However, FDA has expressed concerns about the food safety regulatory systems of some foreign countries. In 2007, FDA issued the *Food Protection Plan*, which sets forth FDA's framework for overseeing food safety, including the safety of imported food;[6] and, at the same time, a 12-agency working group—with FDA participation—issued the Action Plan for Import Safety, which contains, among other things, recommendations for improving the safety of food imports entering the United States.[7] According to the *Food Protection Plan*, while many foreign countries have well-developed regulatory systems to ensure food safety, other countries have systems that are less well developed and that may not be able to ensure food safety to the same degree.

The Federal Food, Drug, and Cosmetic Act does not require FDA to consider where food ingredients are manufactured and the agency does not require companies to identify this information as part of their GRAS determinations, whether they submit that determination to the agency or not. As a result, FDA is not informed about the nature or extent of foreign GRAS substances in the nation's food supply—notwithstanding its concerns about the food safety regulatory systems of some foreign countries, as expressed in the *Food Protection Plan*. However, FDA officials noted that if a concern arises about the safety of an imported GRAS substance, the agency could take enforcement action, such as requiring that the product be brought into compliance, destroyed, or reexported. FDA has taken action on imported GRAS substances, including stevia.

FDA Has Not Taken Steps that Could Help Ensure the Safety of GRAS Determinations

While FDA has issued guidance to minimize the potential for conflicts of interest among its own staff who look at scientific issues and the safety of GRAS substances, it has not issued any guidance on the subject for companies to use with their own scientific experts. FDA has a number of guidelines and policies to ensure that FDA employees, including those who serve on the agency's GRAS Notice Review Teams, as well as individuals who serve on agency scientific and advisory panels, are free from financial conflicts of interest.[8] These federal guidelines, however, do not extend to expert panels convened by private companies to establish consensus for GRAS determinations. In determining whether a substance is GRAS, companies must show that there is common knowledge among qualified experts about the safety of the GRAS substance. According to the 1997 proposal, companies can demonstrate this consensus in a variety of ways, such as assembling scientific review articles, convening a panel of experts, or using reports from authoritative bodies, such as the National Academies. These approaches can also be used in combination. Expert panels can be comprised of a company's own staff or outside experts hired by the company or by a consulting firm. According to FDA officials, it is not uncommon for companies to use expert panels to demonstrate that there is a consensus regarding the safety of their GRAS substance. However, FDA has not issued any conflict of interest guidance that companies can use to help ensure that the members of their expert panels are independent in their determinations of GRAS status. Furthermore, FDA does not call for any information or assurance from companies in their GRAS notices regarding their expert panelists' independence and potential conflicts of interest; thus, FDA does not know whether the determinations of companies' expert panels are arrived at independently.

Scientific, industry, and consumer group officials have raised concerns about the potential for conflicts of interest among members of expert panels used by companies in making GRAS determinations. For example, two food scientists noted that there is a relatively small community of experts qualified to sit on these panels and, inevitably, these experts may have corporate or financial affiliations that could bias their decisions. These officials also said FDA should issue conflict of interest guidelines for expert panels as a way to minimize bias and promote transparency. Similarly, an industry consultant stated that experts who serve on panels come from narrow fields of science and may have developed some of the information that the panels are assessing. In another case, officials from a consumer group questioned whether company GRAS determinations are based on independent scientific evaluations, noting that companies can create an expert panel from either their own staff or from individuals they hire. Furthermore, a food industry official indicated that although this official's company had developed its own conflict of interest guidelines for expert panels, FDA's issuance of conflict of interest guidelines for company use would, among other things, create consistent definitions of expert and independence. Finally, while an official from a consulting firm that convenes expert panels for GRAS manufacturers was confident in the independence of the experts employed by his firm, this official acknowledged that the experts chosen were not asked to complete financial disclosure statements or otherwise provide information on their financial investments.

FDA officials stated that while the agency has the statutory authority to develop guidance for companies' expert panels, FDA officials do not know of any generally available industry guidelines that companies could draw upon in ensuring the independence of their expert panels. Companies may seek to avoid such conflicts on their own. For example, officials of one company stated that they seek to ensure independence by determining if members of their expert panels have any contractual ties that might conflict with their responsibilities. FDA, however, does not have any information on what steps companies take to ensure independence. FDA officials explained that, as a counterbalance to any potential conflicts of interest among companies' experts, the agency's review of GRAS notices does not depend entirely on the conclusions of the expert panels—the agency also considers other available information, such as scientific review articles or the opinions of authoritative bodies. However, while this step may apply to GRAS notices submitted to the agency, it would not apply to GRAS determinations that were not submitted to FDA's voluntary notification program.

In addition, FDA has not taken certain steps to ensure companies maintain proper documentation to support their GRAS determinations. FDA has indicated that it would take steps to help ensure that GRAS determinations were arrived at soundly and that appropriate documentation was maintained. In its 1997 proposal, FDA stated that it would be prudent for companies, including those participating in the voluntary notification program, to maintain documentation of their GRAS determinations and for FDA to monitor compliance with the essence of the statutory requirement—that there is common knowledge among qualified experts that there is reasonable certainty that the GRAS substance is not harmful under the intended conditions of use. Accordingly, FDA announced in the 1997 proposal that it intended to conduct random audits of data and information maintained by these companies. However, according to FDA officials, the agency has not conducted such audits. Agency officials explained that, instead, they have decided to ask for additional supporting documentation only when they determine it is needed.

FDA has not addressed appropriate levels of documentation for companies that do not notify the agency of their GRAS determinations, either in the 1997 proposed rule or in any guidance. To conduct random audits of these companies' GRAS determinations, FDA would need to require them to inform the agency of those determinations. FDA officials stated that companies making GRAS determinations without notifying FDA were not a concern because the Federal Food, Drug, and Cosmetic Act makes companies, not FDA, responsible for GRAS determinations.

Agency officials maintained that they would take enforcement action against any company that had inappropriately determined a substance to be GRAS. However, the possibility of random audits of supporting documentation would provide an added incentive for companies to conduct GRAS determinations appropriately. Without random audits of all companies that make GRAS determinations, FDA has less assurance that these companies have conducted these determinations appropriately, including appropriately documenting the determination and maintaining this documentation.

Table 3. Comparison of the Transparency of FDA's Voluntary Notification Program, Company GRAS Determinations without Notification to FDA, and the Flavor and Extract Manufacturers Association's GRAS Determinations.

GRAS process	FDA is informed about GRAS determinations	FDA is informed about scientific basis of GRAS determination	Information about GRAS determinations publicly available	Number of GRAS substances
FDA voluntary notification program	Yes	Yes	Yes	274[a]
Company GRAS determination without notification to FDA	No	No	No	Unknown
Flavor and Extract Manufacturers Association GRAS determination	Yes	Yes	Yes	2,648[b]

Source: GAO analysis of FDA and Flavor and Extract Manufacturers Association data.
[a]For the period 1998 through 2008. [b]For the period 1960 through June 2009.

Lastly, finalizing the 1997 proposed rule, which FDA considers interim policy, would firmly establish the framework and criteria for FDA's voluntary notification program. It would also reduce the inherent uncertainties for companies of working with an interim policy. For example, FDA could clarify changes it has made in its time frame for completing reviews. Furthermore, according to representatives from the Grocery Manufacturers Association, issuing a final rule would bring more credibility to the voluntary notification program. In addition, FDA has not yet responded to public comments on the proposed rule from over 30 organizations. For example, a consumer group—the Center for Science in the Public Interest—recommended in a 1997 comment that FDA seek authority from Congress to require companies to inform FDA of all GRAS determinations they make. According to FDA officials, while the agency plans to issue a final rule, the agency has had higher priorities and currently has no specific schedule for doing so. However, these officials also said that the program has been operating effectively under the proposed rule.

A Trade Association Informs FDA of Its GRAS Determinations, Even Though It Does Not Participate in the Agency's Voluntary Notification Program

Actions taken by the Flavor and Extract Manufacturers Association help FDA better ensure the independence of scientific assessments of the association's GRAS determinations and obtain information about these determinations. This association conducts GRAS determinations exclusively for its approximately 70 member companies that manufacture these substances. Once a member company submits a flavor or extract—known as a flavoring substance—to the association's GRAS process, the company is not supposed to market it until the association determines the substance is GRAS. To conduct its GRAS determinations, association staff first assess whether the substance will likely meet the criteria for a GRAS determination and whether additional support is needed. When the staff determine that they have sufficient information, they submit the substance to the association's own expert panel, a standing panel of eight academic experts. In hiring panelists, the association requires that they complete a financial conflict of interest form. To further avoid the potential for conflicts of interest, panelists do not know which company has submitted a substance and do not have any contact with applicants' representatives regarding individual substances. Once the expert panel has completed its review, it determines if the substance is GRAS. Association members generally do not seek review through FDA's voluntary notification program, instead relying on the integrity and credibility of the association's process to ensure the marketability of their GRAS substances.

In addition, the Flavor and Extract Manufacturers Association voluntarily informs FDA of its GRAS determinations, including the name of the substance, its properties, and the basis of the determination. According to association officials, the association has provided such information to FDA on all of its GRAS determinations—over 2,600 since 1960. In addition, the association has published journal articles on the workings of its expert panel. It also announces its GRAS determinations in a food industry trade magazine and makes these publications available on the association's Web site for a fee. As table 3 shows, the association's GRAS process achieves a level of public disclosure and agency notification similar to FDA's voluntary notification program.

FDA incorporates the information provided by the Flavor and Extract Manufacturers Association about its GRAS substances into the agency's toxicological database,[9] as well as into EAFUS. FDA officials said that the agency uses the information in these databases to enhance its understanding of the range and amount of GRAS substances likely to be ingested by the public, as well as individual substances' toxicological profiles. FDA would otherwise have to develop some of this information at its own cost. More generally, the information provided by the association enables FDA to be better informed about the composition of the food supply.

FDA officials stated that the agency would have to seek authority from Congress in order to require all companies to inform it of their GRAS determinations. These officials also expressed concern about the potential burden on companies and the availability of resources at FDA to process and assimilate this information in its databases. However, they also said that receiving such information may have the potential to provide additional food safety protection and would allow FDA to be more fully informed about food in the marketplace, including GRAS determinations made by foreign companies in countries with less stringent food safety standards that may pose a threat to the U.S. food supply. These officials added that if the provision of such information was made mandatory, it would be important for FDA to implement this requirement efficiently to emphasize the provision of only information that will be useful to the agency, such as information on novel applications of substances in food. We note that this focused approach, along with implementing this change moving forward rather than retrospectively, as well, could limit the burden of such a requirement on companies and FDA.

FDA Is Not Systematically Ensuring the Continued Safety of Current GRAS Substances

FDA does not systematically reconsider the safety of GRAS substances as new information or new methods for evaluating safety become available. In addition, FDA is generally unaware of companies' reconsiderations of their GRAS determinations because companies are not required to share information about their reviews of the continued safety of their GRAS substances.

FDA Does Not Systematically Reconsider the Safety of GRAS Substances as New Information Becomes Available

The GRAS status of a substance can change and must be reconsidered as new information comes to light or new methods of evaluating its safety arise, according to FDA regulations. The GRAS status of a substance is subject to review as new scientific information is developed that raises questions about the substance's continued safe use. FDA may also consider whether specific information brought to the agency's attention through routine correspondence from interested parties or through a citizen petition raises such safety questions. If FDA decides to review a substance's GRAS status, it may advise companies and other interested parties of those questions by letter.

FDA last engaged in a systematic reconsideration of the safety of GRAS substances in the 1970s and 1980s. This effort raised questions about the safety of almost three dozen GRAS substances. FDA undertook this reconsideration because, during the late 1960s, new scientific information raised questions about the safety of cyclamate salts, a class of artificial sweeteners previously considered GRAS. FDA decided to evaluate, by contemporary standards, the available safety information related to substances it considered GRAS. If the evaluation confirmed that the use of a particular substance was GRAS, FDA issued a new regulation affirming that finding.

To conduct this systematic reconsideration, FDA contracted with an independent scientific organization—the Federation of American Societies for Experimental Biology, which set up the Select Committee on GRAS Substances—to systematically evaluate ingredients considered GRAS at the time. FDA asked the committee to recommend any restrictions that the agency should place on the use of the substances to ensure their safe use in food. Over 10 years—from about 1972 through 1982—the committee reviewed the safety of 422 substances directly added to food and transmitted reports on these substances to

FDA. In all, the committee questioned the safety of 35 of these substances. For 30 of these substances, the committee reported that, unless evidence was provided to FDA showing these substances' safety, it expected FDA to revoke their GRAS status. For example, the committee reported that it could find no information in the scientific literature regarding carnauba wax—a substance of plant origin used at a low level in food products since 1900—and that it, thus, had insufficient data upon which to evaluate its safety. For the remaining 5 substances, the committee found that the current evidence did not show the substance was not harmful at current levels of consumption. For example, the committee reached this finding in examining sodium chloride, or salt, and suggested the development of guidelines for restricting the amount of salt in processed foods and labeling the sodium content of foods. See appendix I for additional information on the GRAS status of salt. As of December 2009, FDA had affirmed 17 of these 35 substances as GRAS by issuing regulations, including a regulation for carnauba wax. FDA had not issued regulations on the remaining 18 substances and could not readily explain why, even though almost 30 years had passed since the committee completed its work. FDA has not revoked the GRAS status of any of these 18 substances whose safety the committee questioned.

Since 1982, FDA has not systematically reconsidered the safety of substances considered to be GRAS as new scientific information has come to light. Specifically, the agency has not contracted for or performed any comprehensive reviews of substances considered to be GRAS and has not developed a formal approach for reviewing these substances. Agency officials stated that they use a database called the Priority-Based Assessment of Food Additives—which contains administrative, chemical, and toxicological information about food ingredients, including GRAS substances—to help prioritize substances for assessment. However, FDA officials could not provide any examples of a reconsideration of the safety of a GRAS substance that resulted from their use of this database. FDA officials also said they keep up with new developments in the scientific literature as part of their professional responsibility as scientists and, on a case-by-case basis, information brought to the agency's attention could prompt it to reconsider the safety of a GRAS substance. Specifically, FDA officials said they may become aware of safety concerns related to GRAS substances through other means, such as through reports in the media or trade press; informal inquiries or complaints from consumers, interest groups, or companies; citizen petitions; or reports published by authoritative bodies. For example, FDA officials stated they are reviewing the issue of companies adding caffeine, a GRAS substance, to certain products, such as alcoholic beverages, after becoming aware of the practice through media reports and from other sources.

Concerns about the safety of certain GRAS substances have led to changes in their GRAS status in the past. According to FDA officials, the agency has not revoked the GRAS status of any substance approved through the petition affirmation process that began in 1972 or retracted its no questions letter for any GRAS notice receiving that agency response since 1997. Nevertheless, questions about the safety of an ingredient previously considered to be GRAS, and changes to that GRAS status, have occurred. In addition to banning cyclamate salts in 1969, other examples of FDA action on the status of GRAS substances include the following:

- In 1985, FDA banned cinnamyl anthranilate, a flavoring agent that had been previously considered GRAS, after studies linked it to liver cancer in mice.
- In 1986, FDA prohibited the use of sulfites, considered GRAS since 1959, on fresh fruits and vegetables intended to be served raw because of potentially severe allergic reactions among those with a sulfite sensitivity; the agency also implemented labeling requirements for other foods containing any added sulfites.

In another more recent case, studies have raised health concerns about the trans fats in partially hydrogenated vegetable oils, and several government and scientific organizations have recommended minimizing consumption of trans fats. FDA has not, however, revoked the GRAS status of these oils. Rather, agency officials indicated that, in response to a 2004 citizen petition, FDA set up a review team in 2004 that is actively reviewing the oils' GRAS status and plans to issue its findings in 2010.[10] See appendix I for additional information on the GRAS status of these oils.

FDA officials told us that information brought to the agency's attention could prompt the agency to reconsider the safety of a GRAS substance. However, we found that FDA has largely not responded to the concerns that individuals and consumer groups have raised through 11 citizen petitions submitted to the agency between 2004 and 2008. Citizen petitions must be submitted according to a prescribed format and are the most formal path an individual or organization can take to bring a problem to FDA's attention. The agency is to respond to these petitions within 180 days, either indicating its decision or informing the petitioner that the agency has not yet reached a decision.

Nine of the 11 citizen petitions raised specific concerns about the safety of GRAS substances or the way they are used in food. For example, a petition submitted in 2006 cited studies linking diacetyl (a substance used to impart a buttery flavor to processed foods, including microwave popcorn) to severe respiratory reactions and called for FDA to revoke diacetyl's GRAS status. FDA has not yet issued a decision on this petition. As table 4 shows, FDA has not issued a decision on 10 of the 11 petitions submitted between 2004 and 2008. The agency most often cited limited resources and other agency priorities to explain why it had not yet reached a decision on these 10 petitions.

As shown in table 4, in three cases, FDA did not provide evidence that it had sent a 180-day letter. FDA officials indicated that in two of the three cases—those related to carbon monoxide—the agency had no record that it had sent a 180-day letter and stated that this was an oversight on the agency's part. In the third case—the petition on milk protein concentrate—agency officials said a 180-day letter had been sent but they could not find it. A majority of the 180-day letters FDA sent stated that the agency had not reached a decision on the petition because of limited resources and other priorities. In the case of the citizen petition on salt, however, FDA indicated that it needed to collect more information before responding and subsequently held a hearing for interested parties in November 2007. As of December 2009, FDA officials stated that they were still evaluating the comments collected at this hearing, among other actions (for information on additional actions FDA is taking related to salt, see appendix I).

FDA has reviewed some of the 10 unanswered citizen petitions more intensively than others. For several of the petitions, FDA officials provided documentation showing extensive review of the concerns raised. For example, in response to one of these petitions, FDA's review included contracting for an independent evaluation of published literature and a calculation of the intake estimates for dietary exposure to the substance. FDA appears to have extensively reviewed the information needed to respond to the citizen petition in this and other cases, but had still not responded to the petitions. In two of these cases, FDA officials had developed internal memoranda recommending a particular response, but the agency had not yet finalized its responses as of December 2009, and agency officials cautioned that final decisions were not necessarily imminent. On the other hand, FDA appears to have only minimally reviewed three of the citizen petitions—those on milk protein concentrate, carbon monoxide in the packaging of fresh tuna, and iodized salt. The agency did not provide documentation to show that any review of these three petitions had occurred.

Table 4. Status of FDA Response to Citizen Petitions on GRAS Substances Filed from 2004 through 2008.

Subject of citizen petition	Concerns raised	Date filed	180-day letter sent	Decision
Milk protein concentrate	Lack of evidence for GRAS status	4/28/2004	No	Pending
Partially hydroge-nated vegetable oils	Increased risk of coronary heart disease from trans fats	5/18/2004	12/21/2004	Pending
Aluminum-based food additives	Link to Alzheimer's disease and elderly cognitive impairment	9/14/2005	3/13/2006	Pending
Salt	Risk of elevated blood pressure from excess consumption of salt	11/8/2005	6/5/2006	Pending
Carbon monoxide gas in fresh meat packaging	Consumer deception and food safety risks	11/15/2005	No	Pending
Carbon monoxide gas in fresh tuna packaging	Consumer deception and food safety risks	3/16/2006	No	Pending
Diacetyl	Lung disease and impairment from inhalation of the substance	9/12/2006	3/6/2007	Pending
Iodized salt	Lack of information on food ingredient Labels	5/7/2007	11/2/2007	Pending
Monosodium glutamate	Substance's links to rise in obesity, diabetes, and autism	12/28/2007	7/18/2008	Pending
Carrageenan and similar substances	Harmful effects on human intestinal cells	6/11/2008	12/9/2008	Pending
Stevia extracts	Therapeutic uses of the substances and questions about their safety	10/7/2008	12/16/2008	Petition denied

Source: GAO analysis of FDA data.

Note: Information in the table is as of December 2009.

While some of these citizen petitions may help FDA better ensure the safety of GRAS substances, GRAS determinations made without notice to the agency remain outside a third party's independent evaluation. Others—including academic experts, consumer groups, and scientific organizations—can play a useful role in helping FDA oversee the safety of GRAS ingredients. In addition to the independent reconsideration of the safety of GRAS substances FDA contracted for in the 1970s and 1980s, independent scientific research has also contributed to FDA's reconsideration of GRAS substances in the past. However, without knowledge of companies' GRAS determinations, third parties, such as the ones that have filed citizen petitions in the past, do not have the opportunity to investigate the potential health effects of such GRAS substances, leaving an additional gap in the oversight of their continued safety.

FDA Is Generally Not Aware of Companies' Reconsiderations of their GRAS Determinations

FDA generally does not know to what extent, or even whether, companies track the evolving scientific information regarding substances the companies have determined are GRAS. Food companies are responsible for ensuring they market safe food, including ensuring the continued safety of the GRAS substances they use as new scientific information emerges. However, companies are not required to share information about their reviews of the continued safety of GRAS substances with FDA. When companies determine a substance is GRAS without notifying FDA, they are the only ones who can ensure the continued safety of that GRAS substance. Representatives of one company told us that they review the status of their GRAS ingredients and keep up with the scientific literature, although they do not generally share the findings of such reviews with FDA. However, FDA officials stated that, in some cases, companies do share with the agency updated scientific information on GRAS substances that were previously reviewed by FDA. As an example, FDA officials stated that, in the past, industry representatives had provided the agency information favorable to the safety of monosodium glutamate, marketed as a flavor enhancer.

FDA is aware of some reconsiderations conducted by companies because the Flavor and Extract Manufacturers Association periodically reconsiders the thousands of substances it has determined to be GRAS and publishes the results of these reviews. According to the association, its GRAS assessment process incorporates new information as it becomes available. In fact, the association's expert panel periodically conducts comprehensive and systematic reviews of all

GRAS flavoring substances that its members manufacture and reviews any individual substances for which potentially significant new data become available. The expert panel conducted its first comprehensive review from 1965 to 1985, evaluating all available new data on the approximately 1,200 substances it had previously determined to be GRAS. The panel affirmed the GRAS status of almost all of the GRAS flavoring substances, but did revoke GRAS status for three. The expert panel conducted a second systematic review between 1994 and 2005 and reviewed all available information relevant to the safety assessment and GRAS status of the approximately 2,000 flavoring substances it had designated as GRAS. This second review process did not result in the revocation of GRAS status for any flavoring substance. In 2009, the association began its third comprehensive review, focusing on flavoring substances in certain structural classes that showed a significant increase in the association's 2005 survey. The results of this review will, as with previous reviews, be published in the peer-reviewed scientific literature, according to an association representative. In addition to these comprehensive reviews, the panel periodically becomes aware of significant new data on prior GRAS decisions during its review of the available scientific data related to flavoring substances. In these cases, the panel re-evaluates the safety of the flavoring substance and may conclude the substance is no longer GRAS. In some instances, the expert panel requests that additional studies be performed by industry members. Over the last four decades, these two review processes have led to numerous studies to address a variety of safety assessment issues that arose during the reviews. Most of these studies have been published. The two review processes also resulted in 11 substances being removed from the association's list of GRAS flavoring substances.

FDA's Regulatory Approach Allows Engineered Nanomaterials to Enter the Food Supply without the Agency's Knowledge

Nanotechnology presents potential challenges to the regulation of food safety, especially because companies may conclude that their engineered nanomaterials are GRAS without informing FDA. FDA has issued some guidance to companies regarding applications of nanotechnology in food. In Canada and the European Union, any such engineered nanomaterials are required to undergo review by government regulators before they can be marketed.

Nanotechnology Presents Potential Challenges to the Regulation of Food Safety

Nanotechnology has many potentially beneficial uses in food. For example, engineered nanomaterials could be used to monitor food quality and freshness; improve the traceability of food products (the ability to track these products from point of origin to retail sale); and modify the taste, texture, and fat content of food. However, the largest area of current usage appears to be in food packaging, where applications such as antimicrobial nanofilms—thin layers of substances meant to hamper the growth of bacteria and fungi—may help bolster food safety.

While applications of nanotechnology with potential food safety benefits have been proposed, reports issued or commissioned by FDA and foreign food safety agencies have identified a number of challenges to the regulation of engineered nanomaterials in food. Specifically, in 2007 an FDA taskforce reported on how nanotechnology might affect the products the agency regulates, including its potential applications in food. The taskforce concluded that the use of engineered nanomaterials presents several challenges, including ensuring the adequacy of methods for evaluating the safety of these engineered nanomaterials in food. The report highlighted the shortcomings of FDA's knowledge about the use of nanotechnology in food, such as the difficulties in identifying its use and in extrapolating natural-scale safety information to nanoscale materials. The taskforce specifically refrained from defining nanotechnology, stating that a definition would be premature given the current limited knowledge. In December 2009, FDA officials said that although the agency's scientific understanding of nanotechnology continues to evolve, in their view, the principles expressed in the taskforce's 2007 report are sound.

Similarly, a Canadian expert panel gathered at the request of the Minister of Health reported in September 2008 that the scientific knowledge needed to assess the risks associated with engineered nanomaterials is limited, especially given the diversity of these materials and their potential applications. The panel found that (1) nanomaterials can pose particular challenges to risk assessment and, hence, to regulation, because they exhibit properties based on both their physical structure and their chemistry; (2) while human and ecological risk assessment frameworks are robust, their application to nanomaterials requires new ways of measuring exposure, dose, and response; and (3) data are inadequate for informing quantitative risk assessments on current and emerging nanomaterials. Because of the limited state of scientific knowledge regarding many nanomaterials, the expert panel stated that priority should be given to a strategic research agenda to improve the understanding of the risks associated with different types of nanomaterials.

The panel also found that high priority should be given to research on how to measure and detect the presence of nanomaterials, nanomaterials' properties that are linked to biological responses, and effective monitoring and surveillance strategies.

Finally, in a February 2009 scientific opinion on nanotechnology and food, the European Union's Food Safety Authority concluded that several challenges still must be addressed in order to ensure the safe inclusion of engineered nanomaterials in food. It recommended further development of risk assessment and safety evaluation methods. The opinion emphasized that, although case-by-case evaluation of specific engineered nanomaterials may currently be possible, risk assessment processes are still under development for characterizing and analyzing these materials in food, optimizing methods to test their toxicity, and interpreting the resulting data. It also stated that there may be additional toxic effects caused by engineered nanomaterials that are not readily detectable by current standard protocols. The opinion concluded that, under these circumstances, any individual risk assessment is likely to be subject to a high degree of uncertainty until more data on and experience with testing of engineered nanomaterials become available.

Companies May Market Engineered Nanomaterials as GRAS without Informing FDA

Despite the challenges inherent in assessing the safety of food ingredients containing engineered nanomaterials, under the Federal Food, Drug, and Cosmetic Act and FDA regulations, a company may market such an ingredient without informing FDA as long as the company has concluded the substance is GRAS. FDA's nanotechnology taskforce began its regulatory policy inquiry by reviewing the agency's authorities to meet any unique challenges that may be presented by FDA-regulated products containing nanoscale materials. The taskforce recognized that, although FDA's authorities may be adequate to meet these challenges, in some cases the evolving state of the science regarding nanotechnology may warrant a case-by-case approach to assess whether sufficient evidence exists to show that products satisfy the applicable statutory and regulatory standards. After reviewing the uncertainties associated with the safety of food ingredients containing engineered nanomaterials, FDA has decided that, at this time, it does not need additional authority to regulate such products, nor does it need to significantly alter its regulatory approach.

FDA has, instead, encouraged companies considering using nanomaterials in food and food packaging to consult with the agency about which regulatory track to follow, including whether such a substance might be GRAS.[11] In these presubmission meetings, companies may discuss this and other issues relevant to their potential submission with FDA officials. FDA officials explained that they were wary of a "one size fits all" regulatory approach for food substances containing engineered nanomaterials. They also stated that some substances that are GRAS at their natural scale may still be GRAS if they were engineered at the nanoscale, so they do not see a need for changing the agency's approach to GRAS substances at this time. However, others, such as some academic experts and consumer groups, have pointed out that engineered nanomaterials are used specifically because of the novel properties they exhibit at the nanoscale; therefore, the fact that a substance is GRAS at its natural scale may not be a good indicator that the new properties the substance takes on at the nanoscale are safe. Nevertheless, the decision to notify FDA of a GRAS substance, even one that contains engineered nanomaterials, is still voluntary.

While FDA officials said that allowing companies to voluntarily provide information about the use of engineered nanomaterials in GRAS substances is sufficient to ensure food safety, few companies participated in another federal agency's voluntary program to gather information about applications of nanotechnology in products. In 2006, the Environmental Protection Agency (EPA) began developing the Nanoscale Materials Stewardship Program, which sought to build the capacity of the agency to deal with materials at the nanoscale. The program consisted of two parts: asking companies to voluntarily (1) supply existing information about their products and (2) conduct further studies to produce new information about their products. According to EPA's interim report, the program suffered from underreporting on the part of nanotechnology manufacturers. EPA reached this conclusion after comparing participation in the program with databases compiled by other organizations that list nanomaterials available for commercial and research sale or commercial products for which the manufacturer makes a claim that the product contains nanomaterials. As of September 1, 2008, the program had received information on 106 engineered nanomaterials from 21 companies and associations. Other organizations' databases, however, listed thousands of engineered nanomaterials that companies were advertising as such to potential customers.

The extent to which GRAS substances incorporating engineered nanomaterials have entered the U.S. food supply is unclear. FDA officials indicated that, as of December 2009, no substances that companies described as containing engineered nanomaterials had been submitted to the agency's GRAS notification program. However, the Acting Deputy Director of FDA's Center for Food Safety and Applied Nutrition said that companies have submitted GRAS notices dealing with substances that some might consider engineered nanomaterials, including cyclodextrins—substances used in a number of foods as flavor carriers or protectants, among other things—and synthetic lycopene—an ingredient for use in breakfast cereals, drinks, and several other foods. However, because companies are not specifically required to identify whether substances they submit to FDA contain engineered nanomaterials and GRAS notification is voluntary, FDA has no way of knowing the full extent to which engineered nanomaterials have entered the U.S. food supply in GRAS substances.

FDA does, however, have some information regarding the inclusion of engineered nanomaterials in food contact substances, for which notice to the agency is required prior to marketing. Food contact substances are defined as substances that are intended as components of materials used in manufacturing, packing, packaging, transporting, or holding food if such use is not intended to have any technical effect in such food. From 2007 through September 2009, FDA has had eight presubmission meetings concerning food contact substances that companies have described as incorporating engineered nanomaterials. As a result, FDA has received food contact substance notifications for four of these substances and two—applications of titanium nitride added to a certain kind of plastic—have been approved.

FDA Has Issued Some Guidance to Companies on Nanotechnology in Food

While it did not recommend changes to the agency's regulatory approach, FDA's nanotechnology taskforce did make recommendations that seek to address the regulatory challenges nanomaterials may present, as detailed in table 5. The taskforce stated in its July 2007 report that the steps it recommended would give affected manufacturers and other interested parties timely information about FDA's expectations in order to foster predictability in the agency's regulatory processes. According to the taskforce, this predictability would foster innovation and enhance transparency while protecting public health. Specifically, the taskforce recommended that, for products not subject to premarket authorization, such as GRAS substances, FDA should develop guidance for industry. This guidance should describe what types of additional information companies should include in their GRAS notices submitted to FDA if the products contain engineered nanomaterials. The task force also recommended that FDA issue a notice in the *Federal Register* requesting that companies voluntarily provide information about their use of engineered nanomaterials in such products.

Table 5. Selected Actions Recommended by FDA's Nanotechnology Taskforce in July 2007 and Status of Their Implementation.

Recommended action	Status of implementation
Issue guidance to companies regarding identification of the particle size for products not subject to premarket authorization, but for which the company may choose to provide notice (such as a GRAS notification).	Implemented
Issue a notice in the *Federal Register* requesting submission of data and other information addressing the effects on product safety of nanoscale materials in products not subject to premarket authori-zation. The notice would address both new products made with nanoscale materials and existing products that are changed to include or include greater proportions of nanoscale materials.	Implemented
Issue guidance or amend existing guidance to describe what additional or distinct information should be submitted to FDA or generated with regard to the use of nanoscale materials in food ingredients for which a GRAS notification is submitted or the reduction of particle size into the nanoscale range for food ingredients for which an earlier notification had been submitted and not objected to by FDA.	Not implemented
Issue guidance recommending manufacturers consider whether and how the presence of nanoscale materials affects the manufacturing process. Relevant considerations would include both situations when the product contains nanoscale materials and when any part of the manufacturing process involves nanoscale materials, even if those materials do not become part of the finished product.	Not implemented

Source: GAO analysis of FDA information.

As shown in table 5, FDA has implemented some, though not all, of the recommendations made by its taskforce and endorsed by the FDA Commissioner in 2007. FDA has updated three of its chemistry guidance documents—for food contact substances, issued in 2007; and for direct food additives and color additives, both issued in 2009—to include guidance regarding identification of the particle size in the substances. FDA officials indicated that the GRAS voluntary notification program relies on the chemistry guidance for food additive petitions and food contact substances, depending on the substance's use. In August 2008, FDA also issued a notice in the *Federal Register* for a public meeting on nanotechnology and requested data and information addressing the effects on product safety of nanoscale materials in all products, including those not subject to premarket authorization. According to FDA officials, during the public meeting, held in September 2008, FDA repeated its request for the voluntary submission of this type of information to the agency, which was due by October 24, 2008. As of December 2009, FDA officials said that they were still evaluating this information. FDA is in the process of implementing the remainder of the recommendations of the 2007 taskforce report, according to agency officials, though they could not provide a timetable for when these recommendations would be implemented.

Food Ingredients That Incorporate Engineered Nanomaterials Must Be Reviewed in Other Countries Before They Can Be Marketed.

Foreign entities we identified as being particularly active in regulating nanotechnology—Canada and the European Union (which regulates food on behalf of its 27 member countries)—do not have a GRAS exemption that would allow companies to market a food ingredient containing engineered nanomaterials without first notifying and obtaining approval from regulators. According to officials of these entities, all novel foods and food additives they oversee are subject to regulatory review before they are introduced into the market.

The European Union has also taken a step to more directly regulate engineered nanomaterials in food. The European Parliament and the Council of the European Union, the European Union's legislative bodies, recently revised their regulations on food additives. These revised regulations include language clarifying that when there is a change in the particle size of a food additive that has previously been approved, for example through nanotechnology, premarket approval for the altered food additive is required. These regulations took effect in January 2010. The same rules would apply to food contact materials produced through nanotechnology. In contrast, FDA has no similar regulations that would require the agency to review changes in the particle size of a substance being used in food.

The European Parliament and Canadian government are also considering additional steps related to nanomaterials in food, according to European and Canadian officials, respectively. The European Parliament is considering an update to its regulations on novel foods—foods or ingredients that have not been used for human consumption to a significant degree in the European Union prior to May 15, 1997—that, in its draft form, includes measures to regulate engineered nanomaterials in food. Specifically, the proposed update would require that all foods containing engineered nanomaterials undergo premarket authorization. The draft novel food regulation also includes a definition of engineered nanomaterials. In addition, in 2008, the Canadian government developed a new initiative proposing to conduct a survey that would require Canadian importers and manufacturers to report their use of engineered nanomaterials produced or imported in excess of 1 kilogram during the 2008 calendar year. The authority for such a survey would be a one-time request and would not require users to submit information on a continual basis. Canada planned to use this information to aid developing a regulatory framework for nanomaterials and determining which information requirements should best be used for subsequent risk assessment under such framework. Canadian officials stated that they originally hoped to issue this regulation in the spring of 2009, but could not predict, as of November 2009, when they would actually issue it.

Conclusions

One of FDA's principal missions is to ensure the safety of the nation's food supply, but a growing number of substances that companies have determined are GRAS may effectively be excluded from federal oversight. While some view GRAS substances as generally presenting a relatively low risk, questions have been raised about the safety of numerous GRAS substances over the last 50 years, and some have been banned as a result. In the future, other substances now considered GRAS may also prove to be unsafe. However, FDA may be constrained in detecting any such future problems because it lacks information about an unknown number of substances companies have determined to be GRAS without informing the agency. FDA's public Web site and some of its databases, including its Everything Added to Food in the United States database, are incomplete without information on these GRAS substances. Furthermore, without issuing guidance on how to prevent conflicts of interest and information in companies' GRAS notices regarding expert panelists' independence, FDA has less assurance of the independence of the experts companies employ to support their GRAS determinations. In addition, how companies are to document their GRAS determinations remains unclear and, because FDA does not randomly audit GRAS determinations, FDA has less assurance that companies have conducted and documented their determinations appropriately. Finally, without reconsidering and updating the 1997 proposed rule, FDA may be falling short of fulfilling its food safety responsibilities.

The petition affirmation process and the voluntary notification program have allowed FDA to perform detailed evaluations of companies' GRAS determinations at one point in time. Nonetheless, because FDA now only addresses safety in response to possible concerns that come to its attention, nearly three decades have passed since the agency last systematically reconsidered the safety of all current GRAS substances. Developing a strategy to systematically reconsider the safety of GRAS substances in light of evolving scientific information and methodologies—including allocating sufficient resources to this effort, developing criteria for review, and collecting information on companies' GRAS reconsiderations—would provide better assurance regarding the continued safety of GRAS substances.

Uncertainties persist about how to evaluate the safety of engineered nanomaterials in food. Nevertheless, FDA has only partially implemented its nanotechnology taskforce's 2007 recommendations that the agency issue guidance documents on the use of these materials in GRAS substances, and the agency does not have a schedule for completing the remaining guidance. Moreover, because FDA has not developed a definition of engineered nanomaterials and does not require companies to identify whether their GRAS substances incorporate such materials, the agency may not receive information about the extent to which these materials are being used. Without a strategy to address the potential for engineered nanomaterials to enter the food supply as GRAS substances without the agency's knowledge, FDA may have less oversight over substances whose safety is uncertain.

We recognize there would be some cost to FDA associated with addressing these issues. However, we believe that developing strategies and collecting information to address these issues would cost-effectively contribute to improving the safety of the food supply. For example, FDA has acknowledged the usefulness and cost-effectiveness of the GRAS determination information provided voluntarily by the Flavor and Extract Manufacturers Association. Receiving similar information from other companies on GRAS determinations made outside of the voluntary notification program would likely provide similar benefits.

RECOMMENDATIONS FOR EXECUTIVE ACTION

To better ensure FDA's oversight of the safety of GRAS substances, we recommend that the Commissioner of FDA take the following six actions:

- develop a strategy to require any company that conducts a GRAS determination to provide FDA with basic information—as defined by the agency to allow for adequate oversight—about this determination, such as the substance's identity and intended uses, and to incorporate such information into relevant agency databases and its public Web site;
- develop a strategy to minimize the potential for conflicts of interest in companies' GRAS determinations, including taking steps such as issuing guidance for companies on conflict of interest and requiring information in GRAS notices regarding expert panelists' independence;
- develop a strategy to monitor the appropriateness of companies' GRAS determinations through random audits or some other means, including issuing guidance on how to document GRAS determinations;
- develop a strategy to finalize the rule that governs the voluntary notification program, including taking into account the experience of the program to date, incorporating input from a new public comment period, and reporting to Congress and the public the agency's timeline for making it final;
- develop a strategy to conduct reconsiderations of the safety of GRAS substances in a more systematic manner, including taking steps such as allocating sufficient resources to respond to citizen petitions in a timely manner, developing criteria for the circumstances under which the agency will reconsider the safety of a GRAS substance, and considering how to collect information from companies on their reconsiderations; and
- develop a strategy to help ensure the safety of engineered nanomaterials that companies market as GRAS substances without the agency's knowledge, including taking steps such as issuing guidance recommended by the agency's nanotechnology taskforce, developing an agency definition of engineered nanomaterials, and requiring companies to inform FDA if their GRAS determinations involve engineered nanomaterials.

If FDA determines that it does not have the authority to implement one or more of these recommendations, the agency should seek the authority from Congress.

Agency Comments and Our Evaluation

We provided a draft of this report to FDA for review and comment. In written comments, which are included in appendix IV, FDA generally agreed with the report's findings and recommendations, while raising concerns about certain aspects of several of the recommendations. The agency also stated that it regards the report as an important contribution to its internal deliberations for improving the agency's oversight of all ingredients added to food. For example, FDA noted that the report's recommendations are aimed at strengthening the rigor of independent GRAS determinations, improving FDA's awareness of and oversight of GRAS determinations, and strengthening postmarket oversight of GRAS substances to address new safety concerns, and said it would fully consider the recommendations and other ideas when it moves to finalize the regulation governing the current voluntary GRAS notification program. FDA did not state when it will finalize the regulation.

FDA said it agreed with the first recommendation in the case of a voluntary GRAS submission, noting that it has practices and procedures in place to collect information on the basis for a company's GRAS determination as part of the agency's notification program. However, from a food safety perspective, FDA said it did not agree with a strategy to require any company that conducts a GRAS determination to provide FDA with basic information only and to put such limited information into an agency database or on its public Web site. Specifically, FDA said that, in theory, it would be informative for the agency to have at least an awareness of the existence of substances that are independently determined to be GRAS, even in the absence of a GRAS notice submitted by the company. However, without a regulatory framework that makes notification mandatory, FDA said it cannot ensure that GRAS determinations that are not submitted to the agency's notification program are rigorous, robust, or consistent with the agency's criteria, as outlined in its 1997 GRAS proposal. Thus, FDA indicated that its ability to oversee the safety of added food ingredients, including GRAS substances, would be enhanced if the manufacturer were required, prior to marketing any new substance or new use of an existing substance, to notify FDA and submit scientific evidence demonstrating the safety and legality of the intended use. These comments suggest that FDA would prefer to make notifications mandatory, a step that we agree would allow the agency to better ensure the sufficiency of company GRAS determinations. If this is FDA's intended strategy in response to our recommendation, we encourage the agency to seek legal authority from Congress, as needed, to implement this approach. However, in the meantime, we continue to believe that requiring companies to provide FDA with basic information—as defined by the agency to allow for adequate oversight—on GRAS determinations that are not submitted to the voluntary notification program would be useful to FDA, such as for potential postmarket enforcement actions. We clarified our recommendation to make this clear.

In addition, FDA said that publishing basic information on company determinations that were not submitted to the voluntary notification program might mislead the public into thinking FDA validated these determinations, which could result in an increased use of the substance in food without FDA being aware of the basis for its safety. We believe that if FDA were to post this information on its Web site with an appropriate disclaimer, the public would not be misled into thinking that FDA had validated these determinations. Moreover, publishing this information could be useful to academic experts, consumer groups, scientific organizations, and others, who in turn could alert FDA to information that may be pertinent for evaluating the safety of a GRAS substance.

Regarding our second recommendation, FDA stated that it recognizes that, because the notifier has an inherent interest in the outcome of its GRAS notice, there is the potential for a conflict of interest. To address this concern, the agency noted that GRAS determinations are required to consider the totality of the publicly available information, including potentially unfavorable information. However, FDA said that it could develop nonbinding guidance for convening expert panels as part of GRAS determinations. The agency also said it plans to finalize its GRAS proposal, including the criteria for making and documenting independent GRAS determinations, and would consider the conflicts issue in that rule making. However, because the time frame for finalizing FDA's GRAS proposal is uncertain—the proposal was promulgated in 1997—and companies continue to make GRAS determinations without notifying the agency, we believe that FDA should consider taking additional action until it finalizes its proposal, as an interim step, to minimize the potential for conflicts of interest in these determinations, such as issuing suggested guidance for companies on this issue.

Concerning our third recommendation, FDA noted that in the case of a voluntary GRAS submission, it does not hesitate to ask a notifier to provide certain data or information as an amendment to a GRAS notice. However, the agency said in cases of GRAS determinations that were not submitted to FDA, it has a very limited basis on which to do an audit because it does not know which companies made determinations about particular substances and uses. We note that our recommendation allows for FDA to monitor the appropriateness of such GRAS determinations through means other than random audits. FDA also said that its 1997 GRAS proposal contains extensive information on how to document

GRAS determinations, and that it has posted additional information on its Web site. Nevertheless, the agency indicated it would further consider the documentation issue as it moves toward finalizing the GRAS proposal. Again, we note that the time frame for finalizing this proposal is uncertain. We also note that the information included in FDA's GRAS proposal and on its Web site generally pertains to documenting GRAS determinations that are submitted to FDA's notification program, and not the remainder of GRAS determinations. Furthermore, as discussed, FDA generally has no information on determinations that are not submitted to the agency and, therefore, less assurance that these determinations have been done and documented appropriately. For these reasons, we continue to believe that FDA should also take steps until it finalizes its proposal to better ensure its oversight of the safety of GRAS substances, including issuing guidance on how to document GRAS determinations that are not submitted to the agency's voluntary notification program.

FDA agreed with our fourth recommendation that it finalize its GRAS proposal. The agency indicated that it anticipates reopening the comment period prior to issuance of a final rule. It also said that after analyzing any comments received, it would determine a time frame for finalizing the rule.

Regarding our fifth recommendation, FDA agreed that a system of postmarket oversight for GRAS substances and also for food additives and food contact substances would help to better ensure the safety of the food supply. However, it also said a more comprehensive, sustainable and systematic approach to postmarket review has been hampered by resource constraints; but that it would continue to work on strategies that efficiently use its available resources to mitigate concerns regarding the safety of foods that contain GRAS substances. FDA also agreed that it should develop criteria for circumstances warranting postmarket review of GRAS substances, and that it should allocate sufficient resources to respond to citizen petitions related to GRAS substances in a timely manner. Regarding the latter, FDA noted that an effective strategy would need to include a means of triaging these petitions for their scientific and legal merit, if its limited resources are to be spent wisely. Regarding the potential collection of information on company reconsiderations of the safety of GRAS substances, FDA said that because, under current law, companies are not required to notify FDA of their GRAS determinations, FDA could ask, but not require, companies to provide information on their reconsiderations. While collecting information from companies on their reconsiderations is but one of the steps we suggest FDA take to implement this recommendation, we note that we recommended that, if FDA determines it does not have legal authority to implement a recommendation, the agency should seek this authority from Congress.

FDA agreed with our sixth recommendation that it develop a strategy to help ensure the safety of engineered nanomaterials that companies market as GRAS substances without the agency's knowledge. The agency observed that current scientific uncertainty regarding potential novel properties of nanomaterials and how to test their safety raises questions about the applicability of the GRAS concept to these substances. Accordingly, FDA stated that it will soon issue draft guidance that will help developers of food applications of nanotechnology determine the applicability of this concept. The agency also indicated that it would continue to consider the viability of establishing an FDA-wide or even a foods definition of nanotechnology.

Finally, FDA did not discuss our recommendation that if FDA determines that it does not have the authority to implement one or more of these recommendations, the agency should seek the authority from Congress.

FDA also provided technical comments that we incorporated in the report, as appropriate.

Lisa Shames
Director,
Natural Resources and Environment

APPENDIX I. ADDITIONAL INFORMATION ON SELECTED GENERALLY RECOGNIZED AS SAFE (GRAS) SUBSTANCES

This appendix provides information on three sources' views on two GRAS substances—salt and the trans fats in partially hydrogenated vegetable oils—and the Food and Drug Administration's (FDA) views on the impact of these sources' findings on the GRAS status of each substance. The three sources are the following:

- *Dietary Guidelines for Americans, 2005.*[12] These guidelines have been published jointly every 5 years since 1980 by the U.S. Department of Health and Human Services and the U.S. Department of Agriculture. The guidelines provide authoritative advice for people 2 years of age and older about how good dietary habits can promote health and reduce risk for major chronic diseases. They serve as the basis for Federal food and nutrition education programs.

- *Dietary Reference Intakes.*[13] These are reports developed by the Food and Nutrition Board of the Institute of Medicine, a part of the National Academies, along with Health Canada. These publications provide recommended levels considered safe for consumption of a wide range of nutrients.
- *FDA's Food Advisory Committee.* The committee and its subcommittees provide advice to the FDA Commissioner and others on emerging food safety, food science, nutrition, and other food-related health issues that the FDA considers of primary importance for its food programs. The committee is charged with reviewing and evaluating available data and making recommendations on matters such as those relating to nutrient needs and nutritional adequacy.

Salt

Dietary Guidelines for Americans

Health effects. The guidelines found that, on average, the higher an individual's salt intake, the higher an individual's blood pressure. Decreasing salt intake is advisable to reduce the risk of elevated blood pressure. Keeping blood pressure in the normal range reduces an individual's risk of coronary heart disease, stroke, congestive heart failure, and kidney disease. Many American adults will develop hypertension (high blood pressure) during their lifetime. The guidelines also found that some individuals tend to be more salt sensitive than others, including people with hypertension, African Americans, and middle-aged and older adults.

Recommended limits. The guidelines recommended that individuals consume less than 2.3 grams of sodium per day, or approximately 1 teaspoon of salt (salt is sodium chloride; sodium amounts are discussed because food labels list sodium rather than salt content). Individuals with hypertension, African Americans, and middle-aged and older adults should aim to consume no more than 1.5 grams of sodium per day and meet the potassium recommendation (4.7 grams per day) with food, as shown in table 6.[14] The guidelines also recommended choosing and preparing foods with little salt and consuming potassium-rich foods, such as fruits and vegetables.

Actual consumption. The guidelines found that nearly all Americans consume substantially more salt than they need. On average, the natural salt content of food accounts for only about 10 percent of total intake, while discretionary salt use (salt added at the table or while cooking) provides another 5 to 10 percent of total intake. Approximately 75 percent is derived from salt added by manufacturers. In addition, foods served by food establishments may be high in sodium. Any program for reducing the salt consumption of a population should concentrate primarily on reducing the salt used during food processing and on changes in food selection (for example, more fresh, less processed items; less sodium-dense foods) and preparation.

Table 6. Recommendations from the *Dietary Guidelines for Americans, 2005*, on Daily Levels of Sodium and Salt in the Diet.

Population	(Amounts in grams)	
	Recommended limits of sodium intake	Equivalent amount of salt
General population	<2.3	<5.9
Middle-aged	1.5[a]	3.8[a]
Older adults	1.5[a]	3.8[a]
Individuals with hypertension	1.5[a]	3.8[a]
African Americans	1.5[a]	3.8[a]

Source: GAO analysis of Dietary Guidelines for Americans, 2005 data.
[a]Individuals in these groups should also meet the potassium recommendation (4.7 grams per day) with food.

Dietary Reference Intakes

Health effects. The intakes report found that the major adverse effect of increased salt intake is elevated blood pressure, a risk factor for cardiovascular and renal diseases. On average, blood pressure rises progressively with increased salt intake. Individuals with hypertension, diabetes, and chronic kidney disease, as well as older persons (over age 50) and African Americans, tend to be more sensitive to the blood pressure-raising effects of salt intake than their counterparts. Genetic factors and an individual's diet also influence the blood pressure response to salt.

Table 7. *Dietary Reference Intakes* Recommendations on Daily Levels of Sodium and Salt in the Diet.

Population	(Amounts in grams)			
	Adequate intake for sodium	Equivalent adequate intake for salt	Tolerable upper intake level for sodium	Equivalent tolerable upper intake level for salt
Young adults (age 19-50)	1.5	3.8	2.3	5.9
Older adults (age 50-70)	1.3	3.3	2.3	5.9
Elderly (over age 70)	1.2	3.1	2.3	5.9

Source: GAO analysis of Dietary Reference Intakes for Water, Potassium, Sodium, Chloride, and Sulfate data.

Recommended limits. The report developed an adequate intake and a tolerable upper level intake for sodium.[15] To ensure that the overall diet provides an adequate intake of other important nutrients, for young adults, the adequate intake level was set at 1.5 grams of sodium per day; the daily adequate intake for men and women 50 through 70 years of age was set at 1.3 grams; and for those 71 years of age and older at 1.2 grams (see table 7). For adults, a sodium tolerable upper intake level of 2.3 grams per day was set, although the report noted that this level is not a recommended intake and there is no benefit to consuming levels above the adequate intake. The tolerable upper intake level may well be lower for those groups of individuals who are most sensitive to the blood pressure effects of increased sodium intake (for example, older persons; African Americans; and individuals with hypertension, diabetes, or chronic kidney disease). In contrast, for individuals who are unacclimatized to prolonged physical activity in a hot environment, their needs may exceed the tolerable upper intake level because of sodium sweat losses.

Actual consumption. The report found that it was well-recognized that the current intake of sodium for most individuals in the United States greatly exceeded both the adequate intake and the tolerable upper intake level. Progress in achieving reduced sodium intake will require changes in individual behavior toward salt consumption, replacement of high salt foods with lower salt versions, increased collaboration of the food industry with public health officials, and a broad spectrum of additional research.

Food Advisory Committee

FDA's Food Advisory Committee and its subcommittees did not examine the health effects of salt or sodium consumption in their most recent meetings, which occurred at various times from 2000 through 2005. The Committee and its subcommittees have not met since 2005.

FDA Response to Findings on Salt

We provided FDA with a draft of this appendix and asked the agency to provide its views on the impact of these sources' findings on the GRAS status of salt. In its response, the agency stated that salt is listed as an example of a common food ingredient that is regarded as GRAS in Section 182.1(a) of Title 21 of the U.S. Code of Federal Regulations. In 1982, FDA discussed the GRAS status of uses of salt, opting to defer any change. In 2007, the agency held a public hearing seeking public comment on a citizen petition seeking to reclassify salt as a food additive, requiring reductions in the salt content of processed foods. The citizen petition is actively under review, and FDA has committed to carefully considering approaches to reducing salt and sodium, with particular attention being given to consumer-focused efforts intended to improve food choice in general through public education and potential labeling changes, as well as focusing on cooperative efforts with the food industry to encourage salt reductions in the food supply.

With regard to the findings of the *Dietary Guidelines for Americans* and the *Dietary Reference Intakes*, the agency stated they are important facets of the discussion regarding the safety of dietary sodium. The FDA Food Advisory Committee has not been asked to comment on the safety of salt or its regulatory status; however, FDA has taken another approach to get expert input. FDA is a sponsor of the current effort by the Institute of Medicine to review and make recommendations about various means that could be employed to reduce dietary sodium intake to levels recommended by the *Dietary Guidelines for Americans*. The *Dietary Guidelines for Americans* and the *Dietary Reference Intakes* report have informed FDA activities related to salt and sodium. For example, in a 2007 *Federal Register* notice announcing the "Salt and Sodium Public Hearing," both the *Dietary Guidelines* and *Dietary Reference Intakes* are cited as part of the background discussion regarding dietary sodium leading up to questions about the agency's approach to salt and sodium.

FDA also indicated that it has the authority to revisit salt's GRAS status. The agency has the authority to reevaluate any GRAS ingredient where questions about the safe use of the ingredient may exist. The agency considers GRAS determinations for uses of substances to be dependant on the available science and current evidence of consensus.

The agency also indicated that, since 1982, FDA has primarily relied on labeling policies and regulation, and consumer education to help reduce the sodium content of processed foods and to allow consumers to make more informed choices that will reduce the sodium content of their diets. These labeling policies require sodium content declaration as part of the nutrition facts panel present on most processed foods; require limits on sodium for products labeled as "healthy;" and establish definitions for foods to be labeled as "low," "reduced," or "no" sodium. Additionally, the agency has authorized health claims for foods low in sodium relating to reductions in the risk of heart disease. The agency is now re-examining a variety of approaches for reducing sodium consumption, including, for example, consumer-focused efforts intended to improve food choices and cooperative efforts with the food industry to encourage salt reductions in processed foods.

Regarding whether the agency has the legal authority to regulate the amount of sodium added to or present in foods, FDA indicated that it has the legal authority to regulate the safe use of ingredients, including salt, added to foods in interstate commerce. The agency also noted that other ingredients, such as sodium bicarbonate, are dietary sources of sodium in addition to salt, and regulations promulgated regarding the amount of salt added to foods would not apply to other sodium-containing ingredients, or to sodium naturally present in foods.

In terms of available regulatory options for reducing the addition of sodium to processed foods and restaurant foods, FDA indicated that it could pursue regulatory actions to reduce the addition of sodium to processed foods, depending on the administrative record it is able to assemble. For example, food ingredient regulations could be crafted that would propose to place limits on some uses of salt in processed foods in interstate commerce. In other cases, for example, foods harvested and prepared intrastate (as in some restaurants) would not be within FDA's statutory authority. However, processed foods in interstate commerce used by restaurants are subject to FDA's regulatory authority.

Regarding whether FDA had consulted with the Surgeon General in determining its regulatory approach to salt in foods, the agency stated that reports and information from the Surgeon General serve as important reference materials utilized by the agency, much in the same way that the *Dietary Guidelines for Americans* or the *Dietary Reference Intakes* reports inform agency activity. The

Surgeon General's Report on Nutrition and Health (1988) is often cited as one of the driving forces behind the Nutrition Labeling and Education Act of 1990 that put in place the legal authority for mandatory nutrition labeling (including sodium declaration), as well as most of the other labeling initiatives mentioned previously.

Trans Fats in Partially Hydrogenated Vegetable Oils

Dietary Guidelines for Americans

Health effects. The guidelines found that a high intake of trans fats increases the risk of unhealthy blood lipid levels, which, in turn, may increase the risk of coronary heart disease.

Recommended limits. The guidelines found that, to decrease their risk of elevated low-density lipoprotein cholesterol in the blood, most Americans need to decrease their intake of trans fats. In fact, the guidelines recommended that Americans keep consumption of trans fats as low as possible. They also recommended limiting intake of fats and oils high in trans fats and choosing products low in such fats and oils.

Actual consumption. The guidelines found that, based on 1994-1996 data, the estimated average daily intake of trans fats in the United States was about 2.6 percent of total energy intake. Processed foods and oils provided approximately 80 percent of trans fats in the diet, compared to 20 percent that occur naturally in food from animal sources. Because the trans fats produced in the partial hydrogenation of vegetable oils account for more than 80 percent of total intake, the food industry has an important role in decreasing the trans fat content of the food supply. The guidelines also found that the trans fat content of certain processed foods had decreased and was likely to continue to decrease as the industry reformulated products.

Dietary Reference Intakes

Health effects. The intakes report found that there was a body of evidence suggesting that trans fats increase total cholesterol and low density lipoprotein cholesterol concentrations in blood and, therefore, the risk of coronary heart disease. Because the intake of trans fats and risk of coronary heart disease is a positive linear trend, even very low intakes of trans fats may increase risk.

Recommended limits. The report recommended that trans fat consumption be as low as possible while consuming a nutritionally adequate diet. Because trans fats are not essential and provide no known benefit to human health, no adequate intake or recommended daily allowance was set. Moreover, a tolerable upper intake level was not set for trans fats because the level at which risk begins to increase is very low and a tolerable level cannot be achieved by usual diets and still have adequate intakes of all other required nutrients.

Actual consumption. The report noted that estimating the amount of trans fats in the food supply has been hampered by the lack of an accurate and comprehensive database from which to derive the data and the trend toward the reformulation of products over the past decade to reduce levels. Nevertheless, it also mentioned that early reports suggested a wide range of trans fat intakes, from 2.6 to 12.8 grams per day, and that the most recent estimate placed the mean trans fat intake for the U.S. population aged 3 years and older at 2.6 percent of total energy intake. The report also noted that, while trans fats are present in dairy products and meats, they are high in stick margarine and those foods containing vegetable shortenings that have been subjected to hydrogenation. Examples of foods that contain relatively high levels of trans fats include cakes, pastries, doughnuts, and french fries. Therefore, the intake of trans fats can be reduced without limiting the intake of most essential nutrients by decreasing the serving size and frequency of intake of these foods, or by using oils that have not been hydrogenated.

Food Advisory Committee

The Nutrition Subcommittee of FDA's Food Advisory Committee met in April 2004 and answered two questions posed by the agency related to trans fats:

(1) The Dietary Guidelines Committee may suggest that less than 1 percent of energy should be obtained from trans fats (2 grams per day for a 2,000 calorie diet). Does the scientific evidence support this level?

Members of the subcommittee voted 5-3 to answer the question "no." However, a majority of subcommittee members agreed to transmit a statement to FDA, as an addendum to their response, that although current scientific evidence did not indicate a specific acceptable daily intake for trans fats, the evidence was consistent with reducing trans fat intake to a level of less than 1 percent of energy (2 grams per day for a 2,000 calorie diet).

(2) When compared to saturated fats, are trans fats considered to be more, less, or similarly adverse with respect to coronary heart disease?

The eight members of the subcommittee all voted "yes," that trans fats are more adverse with respect to coronary heart disease. One member of the subcommittee commented that the yes answer did not reflect the considerable uncertainty discussed by the subcommittee as to whether this difference is significant from a public health perspective at the level of trans fat intake that is typical in the United States.

FDA Response to Findings on Trans Fats in Partially Hydrogenated Vegetable Oils

We provided FDA with a draft of this appendix and asked the agency to provide its views on the impact of these sources' findings on the GRAS status of trans fats in partially hydrogenated vegetable oils. In its response, the agency stated that partially hydrogenated oils (which contain trans fats) are considered GRAS based on a history of use prior to 1958. The partial hydrogenation process was developed in the 1930s and has been in widespread commercial use since the 1940s. Two partially hydrogenated oils have been affirmed by FDA as GRAS—partially hydrogenated low erucic acid rapeseed oil and partially hydrogenated menhaden oil. A number of "standards of identity" in FDA's regulations also implicitly permit their use (for example, those for margarine and shortening). FDA is in the process of re-evaluating the GRAS status of partially hydrogenated oils, and has received two citizen petitions on this topic.

With regard to the findings of the *Dietary Guidelines for Americans,* the *Dietary Reference Intakes*, and the FDA Food Advisory Committee, FDA indicated that it is aware of the findings from these sources and will take them into account in its review of the GRAS status of trans fat. Previous versions of the *Dietary Guidelines for Americans* and the *Dietary Reference Intakes* were referenced in the 2003 trans fat labeling rule.

FDA indicated that it has the authority to review and revoke the GRAS status of partially hydrogenated oils that are used as ingredients intentionally added to food. The agency also indicated that it has the legal authority to regulate the amount of trans fat added to foods in interstate commerce. In addition, the agency noted that naturally occurring trans fats (such as those in meat and dairy products) would not be subject to regulatory action.

Regarding what the agency has done to reduce the use of partially hydrogenated vegetable oils and the levels of trans fat in foods, FDA indicated that it published a final rule in 2003 (effective on January 1, 2006) that required that trans fat content be declared in the Nutrition Facts panel of food immediately under the line for saturated fat. As a result of this labeling change, consumers have additional information to help them make heart-healthy dietary choices. In addition, because of trans fat labeling, industry has voluntarily reformulated foods by reducing the levels of added trans fat. FDA is currently evaluating the effect of trans fat labeling on daily intake.

In terms of its available regulatory options, FDA indicated that it could pursue regulatory actions for reducing or eliminating the addition of partially hydrogenated oils and trans fat to processed food depending on what actions the administrative record would support. For example, FDA could propose to revoke the GRAS status of partially hydrogenated vegetable oils, thereby making such oils unapproved food additives. Alternatively, FDA could propose to set limits restricting the amount of trans fat allowed to be added to food in general, or to specific food categories. FDA also could explore whether it can use labeling authorities to provide industry with greater incentive to voluntarily reduce trans fat levels, such as nutrient content claims and disqualifying levels for making health claims. An increase in consumer outreach also is an option to encourage consumers to read food labels and avoid trans fats or choose foods low in trans fat.

Regarding whether FDA had consulted with the Surgeon General in determining its regulatory approach to partially hydrogenated oils and trans fat in foods, the agency stated that reports and information from the Surgeon General serve as important reference materials for the agency. FDA considers all relevant information when making regulatory decisions and would consult with the Surgeon General on this issue as necessary.

Appendix II. Objectives, Scope, and Methodology

The overall objective of this review was to assess the Food and Drug Administration's (FDA) oversight of substances generally recognized as safe (GRAS). Specifically, we assessed the extent to which (1) FDA's oversight of new GRAS determinations helps ensure the safety of these substances; (2) FDA ensures the continued safety of current GRAS substances as new scientific information emerges; and (3) FDA's approach to regulating engineered nanomaterials in GRAS substances helps ensure the safety of the food supply. We also gathered additional information on the safety of two GRAS substances—salt and trans fats in partially hydrogenated vegetable oils—including the views of the *Dietary Guidelines for Americans, Dietary Reference Intakes*, and FDA's Food Advisory Committee; this information is discussed in Appendix I.

To address the extent to which FDA's oversight of GRAS determinations helps ensure the safety of these substances, we reviewed applicable laws and regulations. Specifically, we reviewed relevant portions of the Federal Food, Drug, and Cosmetic Act, including provisions added and modified by the Food Additives Amendment of 1958, and FDA's 1997 proposed rule, which describes the GRAS voluntary notification program. We also reviewed the public comments made on the proposed rule by 32 commenters. In addition, we obtained information on the agency's oversight of GRAS determinations from FDA officials and documentation they provided. We reviewed FDA's 2007 *Food Protection Plan*[16] and the 2007 *Action Plan for Import Safety*[17] issued by a 12-agency working group with FDA participation, and we obtained various FDA guidance documents on conflict of interest. We also obtained information about FDA's Everything Added to Food in the United States (EAFUS) and Priority-Based Assessment of Food Additives databases from FDA's Web site and agency officials. We developed information on the history of FDA's implementation of the GRAS program from information on FDA's Web site and other sources. We gathered information on the 274 GRAS notices submitted to FDA between 1998 and 2008 from FDA's Web site. We then analyzed this information to determine: the number of notices filed, FDA's response to each notice, and the amount of time it took FDA to complete its review of each notice. We compared the number of days it took FDA to respond to a notice to the 180-day time frame the agency has adopted as a goal for completing most reviews. To assess the reliability of the data on GRAS notices from FDA's Web site, we interviewed FDA officials regarding the processes they follow to enter the data and ensure they are complete and accurate. We also reviewed related documentation and examined the data to identify any obvious errors or inconsistencies. Based on this assessment, we concluded that the data from FDA's Web site were sufficiently reliable for the purposes of this report.

To obtain the views of stakeholders on the GRAS process, we interviewed representatives from (1) trade associations, including the Grocery Manufacturers Association and the International Food Additive Council/Calorie Control Council; (2) companies, including a company that had notified FDA of several GRAS determinations and a company we identified as having made a GRAS determination without notifying FDA; (3) consumer groups, including Consumers Union, the Center for Science in the Public Interest, and Food and Water Watch; (4) authoritative bodies, including the Institute of Medicine of the National Academies, the Institute of Food Technologists, and United States Pharmacopeia; and (5) academic institutions, including the George Washington University and the University of Maryland. In addition, we obtained information about the Flavor and Extract Manufacturers Association's GRAS determination process from its representatives and from FDA officials.

To determine the extent to which FDA ensures the continued safety of GRAS substances, we analyzed the 11 citizen petitions regarding GRAS substances that were submitted to FDA during the recent 5-year period from 2004 through 2008, and discussed with FDA officials the agency's response to these petitions. Specifically, we first asked FDA to identify citizen petitions related to GRAS substances; in its response, FDA identified whether these petitions had been closed or were still pending. We then obtained documentation related to the petitions filed between 2004 and 2008, including the petition itself, any FDA correspondence with the petitioner, and documents demonstrating FDA's review of the petition. In addition, we discussed with FDA officials how they ensure the continued safety of GRAS substances, including whether they have revoked the GRAS status of any substance, retracted any no questions letter for any GRAS notice, or taken any other actions affecting the status of GRAS substances. We also reviewed a previous reconsideration of GRAS determinations carried out by the Select Committee on GRAS Substances in the 1970s and 1980s and analyzed FDA's response to the committee's recommendations.[18] Finally, we reviewed documents from the Flavor and Extract Manufacturers Association on its process to ensure the continued safety of its GRAS determinations and discussed this process with association representatives.

To assess whether FDA's approach to regulating engineered nanomaterials as GRAS helps ensure the safety of the food supply, we reviewed FDA's policies related to regulating engineered nanomaterials in food. Specifically, we reviewed FDA's July 2007 nanotechnology taskforce report,[19] which examined how nanotechnology might affect the products the agency regulates, including its potential applications in food; and we gathered information on the status of implementing the taskforce's recommendations from agency officials. In addition, we met with officials at the Environmental Protection Agency to discuss that agency's Nanoscale Materials Stewardship Program and reviewed related documentation. To identify a nonprobability sample of foreign entities that are particularly active in considering regulation of engineered nanomaterials in food, we first spoke with representatives from the Woodrow Wilson International Center for Scholars Project on Emerging Nanotechnologies and an academic expert on nanotechnology from the University of Michigan identified by an official at the National Academies' Institute of Medicine. These experts identified Canada, the United Kingdom, the European Union, and the United Nations as being particularly active on this issue. However, after reviewing documentation from each of these entities and discussions with their officials, we eliminated the

United Kingdom and the United Nations as sources because, while they have taken certain steps to investigate the safety of nanotechnology in food, they are not directly responsible for regulating nanotechnology in food. To examine Canada's efforts in this area, we reviewed a September 2008 study[20] conducted by an expert panel at the request of the Canadian Minister of Health and interviewed officials from Health Canada, the federal department responsible for helping Canadians maintain and improve their health, and the Canadian Food Inspection Agency, the department in charge of the safety of the Canadian food supply. To examine the European Union's efforts, we reviewed the European Food Safety Authority's February 2009 scientific opinion on nanotechnology in food[21] and interviewed officials from the European Union's Directorate General for Health and Consumers and from the European Food Safety Authority. We did not independently verify statements of foreign law. We also collected the views of a variety of stakeholders on engineered nanomaterials in food, including representatives from the Woodrow Wilson International Center for Scholars Project on Emerging Nanotechnologies, the International Center for Technology Assessment, and Food and Water Watch. Finally, we attended the Food and Drug Law Institute's Second Annual Conference on Nanotechnology Law, Regulation, and Policy held in February 2009.

To present more detailed information on two GRAS substances—salt and trans fats in partially hydrogenated vegetable oils—we gathered the views of expert organizations on the safety of these substances. Specifically, we collected the views on these substances expressed in the *Dietary Reference Intakes*, publications that provide recommended levels considered safe for consumption of a wide range of nutrients, issued by the Institute of Medicine of the National Academies; the *Dietary Guidelines for Americans, 2005,* authoritative guidelines that serve as the basis for Federal food and nutrition education programs, issued by the Departments of Agriculture and of Health and Human Services; and the views of FDA's Food Advisory Committee, which provides advice to FDA on emerging food safety, food science, nutrition, and other food-related health issues. We then presented this information to FDA officials to gather their reaction on the impact of these views on the GRAS status of these substances. Our findings are presented in appendix I.

We conducted this performance audit from October 2008 to February 2010, in accordance with generally accepted government auditing standards. Those standards require that we plan and perform the audit to obtain sufficient, appropriate evidence to provide a reasonable basis for our findings and conclusions based on our audit objectives. We believe that the evidence obtained provides a reasonable basis for our findings and conclusions based on our audit objectives.

APPENDIX III. ILLUSTRATION OF OPTIONS AVAILABLE TO A COMPANY SUBMITTING A NOTICE TO THE FOOD AND DRUG ADMINISTRATION'S (FDA) GENERALLY RECOGNIZED AS SAFE (GRAS) VOLUNTARY NOTIFICATION PROGRAM

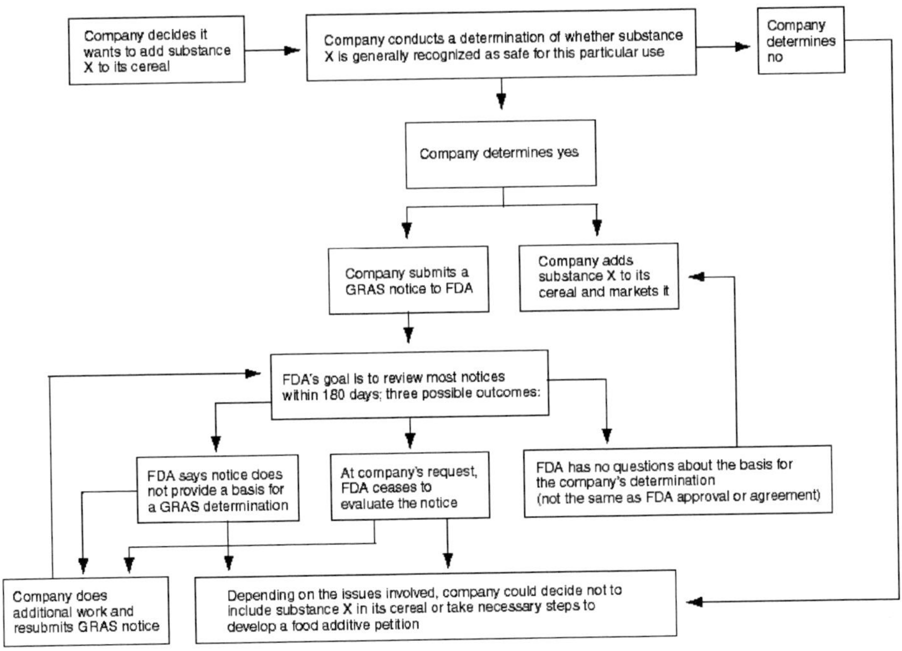

Source: GAO analysis of FDA information.

Notes: (1) This graphic shows some of the steps in FDA's voluntary notification program and options potentially available to a company; it is not intended to show all possible variations. For a more detailed description of the program, see (2) At any point, a company might decide not to include the substance in its product or might proceed to market without FDA's response; if the use of a food substance is GRAS, it is not subject to premarket review and approval by FDA. (3) The analysis assumes the company is acting in accordance with the law; while participation in the notification program is voluntary, companies must comply with the law. (4) Regardless of whether the use of a substance is a food additive use or is GRAS, there must be evidence that the substance is safe under the conditions of its intended use; a GRAS substance is distinguished from a food additive on the basis of the common knowledge about the safety of the substance for its intended use.

Figure 3. Options Available to a Company Participating in the GRAS Voluntary Notification Program.

APPENDIX IV. COMMENTS FROM THE FOOD AND DRUG ADMINISTRATION

DEPARTMENT OF HEALTH & HUMAN SERVICES OFFICE OF THE SECRETARY

Assistant Secretary for Legislation
Washington, DC 20201

JAN 21 2010

Lisa Shames
Director
Natural Resources and Environment
Government Accountability Office
441 G Street NW
Washington, DC 20548

Dear Ms. Shames:

Enclosed are the Department's comments on the U.S. Government Accountability Office's (GAO) draft report entitled: "Food Safety: FDA Should Strengthen Its Oversight of Food Ingredients Determined to Be Generally Recognized as Safe (GRAS) (GAO-10-246).

The Department appreciates the opportunity to review and comment on this report before its publication.

Sincerely,

Andrea Palm
Acting Assistant Secretary for Legislation

Enclosure

GENERAL COMMENTS OF THE DEPARTMENT OF HEALTH AND HUMAN SERVICES (HHS) ON THE GOVERNMENT ACCOUNTABILITY OFFICE'S (GAO) DRAFT REPORT ENTITLED, "FOOD SAFETY: FDA SHOULD STRENGTHEN ITS OVERSIGHT OF FOOD INGREDIENTS DETERMINED TO GENERALLY RECOGNIZED AS SAFE (GRAS)" (GAO-10-246)

Introduction

The Department appreciates the opportunity to review and comment on the Government Accountability Office's (GAO's) draft report. The GAO report identifies important challenges that FDA faces when overseeing the safety of all food ingredients and new food technologies, not just those classified as generally recognized as safe (GRAS) for their intended use. The FDA is committed to exercising all its legal authorities and making efficient use of its resources to ensure the safety of substances added intentionally to food.

In 2009, FDA initiated internal discussions centered on developing a strategy for improving the agency's oversight of all ingredients added to food, either indirectly or directly. Thus, FDA welcomes GAO's review and recommendations and regards the report as an important contribution to FDA's internal deliberations. Any new strategies FDA develops would build on positive aspects of FDA's current pre-market review programs for food ingredients (i.e., food contact substances, food additives, and GRAS substances) and improve the agency's ability to conduct appropriate pre- and post-market reviews of the safety of ingredients added to food, based on current and emerging data and information.

As discussed more fully in the FDA's comments that follow, the Federal Food, Drug, and Cosmetic Act has several criteria that must be met before the intended use of the substance in food can be generally recognized as safe (GRAS): that its safety be established scientifically (or by common use prior to 1958), and that the basis for concluding it is safe must be generally accepted, publicly available, and transparent to experts. The parties producing or using the substance have the ultimate burden of proof with respect to demonstrating its safety for the intended use, and because a GRAS determination is time-dependent, the GRAS status could change as new science or factual data become available.

The law requires new food substances or new uses of substances to be approved by FDA prior to marketing unless the substance is GRAS for its intended use, with sponsors being allowed under the law to make their own independent determination of the GRAS status of a new ingredient or use. As outlined in these comments, however, the same safety standards and public health principles apply to all added food ingredients (other than prior-sanctioned substances), whether FDA approval is required or not.

The GRAS concept was included in the law as a means to avoid wasteful use of government and industry resources in formal approval processes when the safety of an ingredient has already been established and is recognized by qualified experts. The GRAS concept has served well as a source of flexibility to avoid unnecessary FDA reviews of substances with a long history of safe use in food, as well as new substances and uses of substances whose safety is well-established and known.

As the implementation of the GRAS concept has evolved since its enactment in 1958, however, it has become the general rule, rather than the exception, that new substances and new uses of substances enter the market on the basis of the sponsor's independent safety assessment and GRAS determination, rather than through approval by FDA. FDA has established a voluntary mechanism for companies to notify the agency of independent GRAS determinations, but, as the GAO report points out, the inherent nature of the GRAS concept and the evolution in its implementation raise legitimate questions about the scope of FDA's knowledge concerning the uses of food ingredients and the ability of FDA to assess and ensure the rigor of the scientific evaluations underlying independent GRAS determinations.

The existence of these questions impinges on one of the goals that motivated adoption of the law governing food ingredients, namely ensuring public confidence in their safety based on affirmative FDA oversight, and is one of the reasons FDA is reviewing how it oversees food ingredient safety. New food technologies, such as nanotechnology, further challenge FDA's ability to ensure the safety of added food substances.

The GAO's draft report responds to legitimate questions about the GRAS concept and FDA's implementation of it by making several recommendations aimed at strengthening the rigor of independent GRAS determinations, improving FDA's awareness of and oversight of GRAS determinations, and strengthening post-market oversight of GRAS substances to address new safety questions. FDA believes that the GRAS concept has continuing utility as a practical tool for distinguishing between substances and new uses of substances that merit a full pre-market safety evaluation by FDA and those that do not. FDA also agrees broadly, however, that its oversight of GRAS substances and the manner in which independent GRAS determinations are made should be strengthened, and the agency will fully consider GAO's recommendations and other ideas when it moves to finalize the regulation governing the current voluntary GRAS notification program.

FDA is actively considering steps it can take under current law to improve both pre-market and post-market oversight of all added food substances, whether they enter the market based on a GRAS determination or through a formal FDA approval process. In particular, FDA believes that its ability to oversee the safety of added food ingredients, including GRAS substances, would be enhanced if the manufacturer were required, prior to marketing any new substance or new use of an existing substance, to notify FDA and submit scientific evidence demonstrating the safety and legality of the intended use. FDA also agrees with GAO that there is a need for enhancement of post-market oversight, including more effective tools for FDA to obtain from sponsors new information related to the safety of marketed food ingredients so that FDA is better able to identify substances meriting a substantive post-market review. Implementation of these ideas would require additional resources, and some may require new legislation.

Regulatory Framework for Food Ingredients under Section 409 of the Federal Food, Drug, and Cosmetic Act

In 1958, Congress enacted the Food Additives Amendment (the 1958 amendment) to the Federal Food, Drug, and Cosmetic Act (the Act). The 1958 amendment requires that, before a new additive can be used in food, its safety must be demonstrated to the satisfaction of the FDA (i.e., premarket approval). The 1958 amendment, however, also exempted several categories of substances from the definition of a "food additive," including food ingredients that are generally recognized as safe (GRAS) for their intended conditions of use (the GRAS exemption), which is discussed in greater detail below, and food ingredients used in accordance with a "sanction" granted by either FDA or the U.S. Department of Agriculture prior to 1958. Therefore, these exempted food ingredients are not subject to FDA's pre-market review for food additives.

Under section 409 of the Act, if FDA grants a petition for a food additive, it is required to by order establish a regulation that specifies the conditions under which the additive may be safely used. In addition to a regulation, there are other administrative actions pertaining to a food additive petition that involve publications in the *Federal Register*, including the decision to file a petition, deny a petition, stay the effective date of a regulation, respond to objections, respond to a request for a hearing, and amend or repeal a regulation. Rulemaking is also required for FDA to obtain data and information to substantiate the safety of an interim food additive or to remove the additive from the market in light of information establishing a safety concern. FDA must also initiate rulemaking to adopt new food additive specifications of identity and purity or to otherwise amend conditions for the safe use of food substances in its regulations. These administrative procedures for food additives are cumbersome and costly but add little to the quality of FDA's scientific review. Moreover, they provide a disincentive for firms to seek approval of new ingredients as food additives and an incentive for them to make independent GRAS determinations. Thus, a lack of sufficient resources and the time consuming and resource intensive process of rulemaking has often slowed or prevented the FDA from adopting through rulemaking more modern standards even if they improve food safety and are widely implemented by the food industry.

The GRAS Exemption

It is on the basis of the GRAS exemption to the food additive definition that many substances are lawfully added to food without premarket approval by FDA. In establishing the GRAS exemption, Congress determined that many substances intentionally added to food should not require a formal premarket review by FDA to ensure their safety, either because their safety had been established by a long history of use in food or because their safety had been established by the information generally available to, and accepted by, qualified scientists about the substances.

Specifically, under the Act and FDA's regulations, use of a food ingredient may be GRAS through scientific procedures, which include both a technical element and a common knowledge element, as explained in greater detail below. In addition, an ingredient used in food before 1958 may be GRAS instead of through experience that the intended conditions of use are safe based on common use in food.

Under "scientific procedures," a GRAS determination for a particular use of a substance requires both technical evidence of safety (technical element) and a basis to conclude that this technical evidence of safety is generally known and accepted (the common knowledge element).

The technical element of the GRAS standard requires that information about the substance establish that the intended use of the substance is safe. FDA has defined "safe" (21 CFR 170.3(i)) as a reasonable certainty in the minds of competent scientists that the substance is not harmful under its intended conditions of use. The same quality and quantity of data and information are necessary to establish the safety of the use of a GRAS substance as for a food additive use.

The common knowledge element of the GRAS standard includes two criteria: (1) the data and information relied on to establish the technical element must be generally available; and (2) there must be a basis to conclude that there is consensus among qualified experts about the safety of the substance for its intended use. Neither criterion is, by itself, sufficient to satisfy the common knowledge element of the GRAS standard.

As described below, FDA lists a number of food ingredients in its regulations that it considers to be GRAS. In addition to the ingredients listed in FDA's regulations, a company can make an independent determination that the use of a substance is GRAS. If FDA does not agree with an independent GRAS determination, the agency can take enforcement action, including judicial action through the Department of Justice, to stop distribution of the foods containing it on the grounds that such foods are or contain an unlawful food additive.

History of FDA's Approach to the GRAS Provision/Exemption

Following enactment of the 1958 amendment, FDA solicited information on the then current uses of added ingredients that stakeholders considered GRAS with a view toward establishing an initial GRAS list in the Code of Federal Regulations (CFR). The initial GRAS list was included in the CFR in 1961. In addition, during the 1960s, many manufacturers requested FDA's opinion on whether their conclusions of GRAS status were justified and received "opinion letters" from FDA. These "opinion letters" were often only available to the requestor and were formally revoked by FDA in 1970 (21 CFR 170.6; 35 FR 5810; April 9, 1970).

In 1969, FDA removed cyclamate salts from its GRAS list as a result of safety questions, and then-President Nixon directed FDA to reexamine the safety of GRAS substances listed in the CFR. In the 1970s, FDA conducted a comprehensive review of substances presumed to be GRAS. FDA contracted with the Life Sciences Research Office (LSRO) of the Federation of the American Societies for Experimental Biology (FASEB) to summarize the available scientific literature and to recommend restrictions, if any, on the use of the substances needed to ensure their safe use in food. By 1982, the FASEB/LSRO Select Committee on GRAS Substances (SCOGS) had submitted opinions to the FDA on the health aspects of more than 400 substances.

As part of its reexamination of GRAS substances, in the 1970s, FDA developed and continues to maintain and expand a data retrieval and evaluation system to assist in FDA's ability to monitor food ingredients (Priority Based Assessment of Food Additives [PAFA]). Due in part to the increasingly global nature of the food supply chain, the development of novel ingredients and foods and the diversity of available foods, FDA's ability to conduct post-market review and monitoring of food ingredients is a challenge. In response to this challenge, FDA will continue to develop strategies and tools to enable FDA to protect public health through improved oversight of food ingredients.

Also during the 1970s, FDA established a voluntary GRAS affirmation petition process similar to the mandatory food additive petition process. In many cases, those companies that took part in the voluntary GRAS affirmation petition process began to market their products based on FDA's filing of their petition and well before FDA reached a decision on the GRAS status of the petitioned use.

Due to the resource intensive nature of the GRAS affirmation petition process and in the interest of improving efficiencies, in 1997, FDA proposed to replace the voluntary GRAS affirmation petition process with a voluntary notification procedure (GRAS notification program) that directs FDA's resources more effectively to questions about GRAS status that are a priority with respect to public health protection (62 FR 18937; April 17, 1997; Substances Generally Recognized as Safe; the GRAS proposal). To date, FDA has filed over 300 notices.

Challenges Posed by Novel Technologies and Uses

The widespread reliance on the GRAS concept as the basis for market entry of new food ingredients and uses raises challenging issues in the context of novel technologies and novel uses of food ingredients. FDA agrees with the finding in the GAO report that nanotechnology presents particular challenges, since it may involve dramatically reducing the particle size, and in turn altering the physical and biological properties, of familiar GRAS substances without changing the chemical identity of the substance. Similarly, familiar substances are being added to products as varied as infant formula, energy drinks, and alcoholic beverages to produce "novel foods." Overly expansive reliance on the GRAS concept, coupled with the legally available option for companies to make their own GRAS determination without notifying FDA, can undermine FDA's ability to ensure either the safety or legality of novel food ingredients and technologies. The proliferation in the marketplace of products obtained through novel technologies and/or novel uses of food ingredients is another important motivation for FDA's re-examination of the scientific, legal, and regulatory bases for evaluating the implementation of the food additive law.

Response to Recommendations

Under the current legal authorities, a company can either notify FDA through a GRAS notice of its GRAS determination through FDA's voluntary notification program or make a GRAS determination without submitting a GRAS notice to FDA. In terms of GAO's recommendations, five of the six recommendations refer to GRAS determinations generally and are not limited solely to GRAS determinations where a GRAS notice is submitted to FDA. In responding to each of GAO's recommendations, FDA is responding primarily within the context of the agency's current legal authorities and is addressing both types of GRAS determinations, where applicable.

GAO Recommendation 1

To better ensure FDA's oversight of the safety of GRAS substances, GAO recommends that the Commissioner of FDA develop a strategy to require any company that conducts a GRAS determination to provide FDA with basic information about this determination, such as the substance's identity and intended uses, and to incorporate such information into relevant agency databases and its public Web site.

FDA Response to Recommendation 1

In the case of a voluntary GRAS submission, FDA agrees with GAO's first recommendation and has practices and procedures currently in place as part of the GRAS notification program that are consistent with it. When a company voluntarily submits a GRAS notice to FDA, FDA reviews the submission to see that it includes appropriate information before the submission is filed. If lacking important information, such as information about the substance's identity and intended uses, the GRAS notice would not be filed. In such a case, the notifier is contacted and given an opportunity to correct such deficiencies. If the deficiencies are not corrected to FDA's satisfaction, the FDA ceases to review the notification.

Once FDA has completed its review of a GRAS notice and responded to it, information about the notice is posted in FDA's GRAS inventories on the agency's public Web site. Since January 2009, the agency has been publicly posting GRAS notices on its Web site. The Office of Food Additive Safety in CFSAN is currently working on a long-term project to develop a system for data retrieval and evaluation that will incorporate new approaches including computer-assisted analyses; this system will include information from submissions sent to FDA for food contact substances, food additives and GRAS substances (Chemical Evaluation and Risk Estimation System [CERES]).

FDA has also considered how the recommendation to obtain and publicly post basic information about GRAS substances might apply to the situation in which a company makes a GRAS determination for its intended use of a substance and does not notify FDA. This is lawful according to the 1958 amendment to the Act if the use of the substance meets the GRAS criteria. Regardless of whether a company notifies FDA of its GRAS determination, the company is responsible for compliance with the applicable statutes. In addition, FDA expects the criteria for GRAS eligibility and determination through scientific procedures that are outlined in the 1997 GRAS proposal to be met in a firm's GRAS determination. However, without a regulatory framework that makes notifications mandatory, FDA cannot ensure that GRAS determinations that are not currently notified to FDA are rigorous, robust, or consistent with agency's proposed criteria.

In theory, it would be informative for FDA to have, as GAO suggests, at least an awareness of the existence of substances that are independently determined to be GRAS, even in the absence of a GRAS notice submitted by a company. For example, such information could assist FDA with potential post-market enforcement actions. FDA is, however, concerned that, while the legal responsibility for safety and legal compliance would remain with the firm, there might be a public perception or expectation that the listed GRAS determinations were in some way validated by FDA, which could result in an increased use of the substance in food without FDA being aware of the basis for its safety. Thus, from a food safety perspective, FDA does not agree with the recommendation to develop a strategy to require any company that conducts a GRAS determination to provide FDA with basic information **only** and to put such limited information into an agency database or on its public Web site. FDA shares the transparency goal underlying this recommendation, however, and will consider other ways to achieve it.

GAO Recommendation 2

To better ensure FDA's oversight of the safety of GRAS substances, GAO recommends that the Commissioner of FDA develop a strategy to minimize the potential for conflicts of interest in companies' GRAS determinations, including taking steps such as issuing guidance for companies on conflict of interest and requiring information in GRAS notices regarding expert panelists' independence.

FDA Response to Recommendation 2

FDA recognizes that, because the notifier has an inherent interest in the outcome of its GRAS notice, there is the potential for a conflict of interest (this is also true for food additive petitions). To address this concern, GRAS determinations are required to consider the totality of the publically available information, including potentially unfavorable information. In theory, the FDA could develop guidance for convening expert panels as part of GRAS determinations, but this guidance would necessarily be non-binding. As discussed in FDA's response to GAO's Recommendation 4, FDA plans to finalize its GRAS proposal, including the criteria for making and documenting independent GRAS determinations, and will consider the conflicts issue in that rulemaking.

FDA's GRAS rulemaking proposal explains that the bases to demonstrate the common knowledge element of GRAS are varied. These may include publication of data and information in the secondary scientific peer-reviewed literature, documentation of an "expert panel" that is specifically-convened for this purpose; or the opinion or recommendation of an authoritative body such as the National Academy of Sciences. The opinion of an expert panel is useful when multiple studies bearing on the safety of a substance are published but there are no secondary sources that evaluate these studies and draw general conclusions based on this comprehensive body of knowledge. Moreover, the opinion of a specially-convened expert panel can contribute toward providing a basis for showing expert consensus when an individual published study raises safety questions.

There is no requirement to submit the findings of an expert panel to FDA as part of a GRAS notice. When evaluating a GRAS notice containing a report of an expert panel commissioned by a company, the FDA considers the totality of the publically available and corroborative evidence about the safety of the substance for its intended use, including favorable and potentially unfavorable information. FDA considers that the use of an expert panel is one way to demonstrate consensus (i.e., the common knowledge element of safety). FDA does not consider the view of an expert panel alone to be determinative for establishing safety.

When FDA finds that the composition of the expert panel is scientifically deficient in terms of the breadth of scientific expertise or that the data and information considered by the expert panel does not adequately reflect the totality of scientific knowledge, FDA takes this into account during the agency's review or when preparing its response to the GRAS notice.

FDA notes that GAO's report mentions FDA's advisory committees. FDA has a history of using such committees to seek expert scientific advice. The agency has guidance that applies to special government employees and regular government employees (the advisory committee "members") invited to participate in FDA advisory committees subject to the Federal Advisory Committee Act. The agency bears the costs for advisory committees including compensation and travel expenses for members. As such, FDA considers it appropriate to provide guidance that pertains to its advisory committees. In part, the agency issues such guidance to describe the applicable laws and regulations to promote transparency and to ensure the integrity of the deliberations of its advisory committees.

GAO Recommendation 3

To better ensure FDA's oversight of the safety of GRAS substances, GAO recommends that the Commissioner of FDA develop a strategy to monitor the appropriateness of companies' GRAS determinations through random audits or some other means, including issuing guidance on how to document GRAS determinations.

FDA Response to Recommendation 3

In the case of a voluntary GRAS submission, FDA does not hesitate to ask a notifier to provide certain data or information as an amendment to a GRAS notice. Though such requests are not random, the agency uses its resources to seek access to data and information as focused-queries, rather than on a random audit, basis.

FDA has also considered how this recommendation might apply to the situation in which a company does not notify FDA of its GRAS determination. Regardless of whether the company notifies FDA of its GRAS determination, the company is responsible for compliance with the applicable statutes, including the requirement that for a substance to be GRAS there must be common knowledge among qualified experts that there is reasonable certainty that the substance is not harmful under the intended conditions of use. Absent knowledge of which companies have made GRAS determinations about particular substances and uses, however, FDA has a very limited basis on which to audit those determinations.

In terms of how to document GRAS determinations, the preamble of the GRAS proposal provides extensive information, including the elements of the GRAS standard (62 FR 18937 at 18940). Specific to the GRAS notification program, information exists in the GRAS proposal for the documentation of GRAS determinations (62 FR 18937 at 18947).[1] As proposed, the notification procedure would allow FDA to direct its

[1] FDA excerpted a portion of proposed 170.36, and posted this text on the Internet as "How to Submit a GRAS Notice."

resources to questions about GRAS status that are a priority with respect to public health protection. FDA will further consider the documentation issue as it moves toward finalizing the notification proposed rule.

FDA has posted guidance on the agency's website that answers common questions about the food ingredients classified as GRAS in the form of frequently asked questions.[2] This guidance addresses common questions about the regulatory process and regulatory considerations regarding whether the use of a food substance is GRAS.

GAO Recommendation 4

To better ensure FDA's oversight of the safety of GRAS substances, GAO recommends that the Commissioner of FDA develop a strategy to finalize the rule that governs the voluntary notification program, including taking into account the experience of the program to date, incorporating input from a new public comment period, and reporting to Congress and the public the agency's timeline for making it final.

FDA Response to Recommendation 4

FDA agrees with GAO's recommendation to finalize the GRAS proposal on a timeframe that is in keeping with FDA's other public health and rulemaking priorities. The agency anticipates reopening the comment period prior to the issuance of a final rule. After analyzing any comments received, FDA would be able to outline a timeframe for finalizing the GRAS proposal.

GAO Recommendation 5

To better ensure FDA's oversight of the safety of GRAS substances, GAO recommends that the Commissioner of FDA develop a strategy to conduct reconsiderations of the safety of GRAS substances in a more systematic manner, including taking steps such as allocating sufficient resources to respond to citizen petitions in a timely manner, developing criteria for the circumstances under which the agency will reconsider the safety of a GRAS substance, and considering how to collect information from companies on their reconsiderations.

FDA Response to Recommendation 5

FDA agrees that a system of post-market oversight for GRAS substances, and also for food additives and food contact substances, would help to better ensure the safety of the food supply. In fact, FDA has engaged in post-market review of many substances,

---

[2] This guidance is available at http://www.fda.gov/Food/GuidanceComplianceRegulatoryInformation/GuidanceDocuments/FoodIngredientsandPackaging/ucm061846.htm

including GRAS substances, over time as evidenced, for example, by its issuance of numerous Warning Letters for violative substances and by data collection via the Priority-based Assessment of Food Additives (PAFA). In the 1970s, FDA contracted with the FASEB/LSRO Select Committee on GRAS Substances to undertake an exhaustive effort to reevaluate GRAS substances specifically. A more comprehensive, sustainable and systematic approach to post-market review has been hampered by resource constraints. FDA will continue to work on its strategies that efficiently utilize available resources to mitigate concerns regarding the safety of foods on the market that contain added GRAS substances (or new intended uses of GRAS substances) or food additives. FDA notes that post-market oversight involves both review units as well as units that are specialized in enforcement activities.

The FDA also agrees with GAO's recommendation that criteria be developed for the circumstances warranting post-market review of GRAS substances. Generally, post-market reviews are triggered by new or previously unavailable information that suggests adverse effects in animals or man. Such information may point to a real risk and lead to measures taken by FDA or industry that are protective of public health. In other cases, such information may be of uncertain value due to lack of scientific merit or other shortcomings (e.g., the data are contradicted by other data or relate to an area of science that is not well understood). Consequently, FDA agrees that established criteria will help guide FDA in conducting efficient and effective post-market reviews of substances added to marketed foods and that such reviews serve the best interest of public health.

In general, FDA agrees with GAO's recommendation that FDA allocate sufficient resources to respond in a timely manner to Citizen Petitions that relate to GRAS substances as well as food additives, in keeping with other public health priorities. FDA considers Citizen Petitions an important component of post-market review when scientifically sound safety issues are raised and appropriately documented with scientific data and information. Unfortunately, this is not always the case. An effective FDA strategy would need to include a means of triaging Citizen Petitions for their scientific and legal merit if limited FDA resources are to be spent wisely. FDA would need additional resources to ensure timely reviews of Citizen Petitions, various types of regulatory submissions and FDA-initiated reviews (e.g., post-market reviews).

FDA appreciates the intent of GAO's recommendation to collect information from companies on their reconsiderations of the safety of GRAS substances. However, FDA notes that effective implementation of this recommendation is outside of FDA's current authority. The Act does not require companies to notify FDA of GRAS determinations; consequently, under the statute FDA could ask, but could not require, companies to provide information about all GRAS re-determinations.

GAO Recommendation 6

To better ensure FDA's oversight of the safety of GRAS substances, GAO recommends that the Commissioner of FDA develop a strategy to help ensure the safety of engineered nanomaterials that companies market as GRAS substances without the agency's knowledge, including taking steps such as issuing guidance recommended by the agency's nanotechnology taskforce, developing an agency definition of engineered nanomaterials, and requiring companies to inform FDA if their GRAS determinations involve engineered nanomaterials.

FDA Response to Recommendation 6

FDA agrees that it needs to develop a strategy to help ensure the safety of engineered nanomaterials that companies market as GRAS substances without the agency's knowledge. To that end, FDA will soon issue regulatory guidance to developers of nanomaterials describing the agency's thinking regarding how the GRAS concept and other elements of the regulatory framework apply to such materials.

Like all food ingredients, nanomaterials would need to satisfy the statutory standard for GRAS in order to be exempt from FDA's premarket review requirements. From the standpoint of safety evaluation, FDA's concern about engineered nanomaterials is not their size per se. It is rather the fact that, by design, they typically have properties that are different from the conventional-scale versions of the same chemical substance, coupled with the current scientific uncertainty regarding differences in biological interactions between nano- and conventional-scale materials and how to test the safety of the often novel properties associated with nanomaterials. This state of the science suggests that meeting the general recognition standard for GRAS status may be difficult. In some cases the answer may be clear, such as when a nanoscale particle exhibits novel properties potentially affecting its toxicological profile and has not been tested and found safe using validated toxicity testing protocols. In other cases, the analysis may be more complicated, and the answer may change as the science evolves.

FDA has recognized previously that nanotechnology food applications raise difficult scientific and regulatory questions, including in the FDA Nanotechnology Task Force Report published in 2007. As scientific knowledge advances to help FDA better understand technical issues about nanotechnology, FDA's strategy for addressing the safety of nanomaterials used in food and food packaging will continue to evolve. FDA has thus far published three guidance documents dealing with nanomaterials, all of which address with the safety assessment of food ingredients or food packaging components.

FDA is also developing or considering the development of further guidance on the regulatory status and safety of nanomaterials used in food and food packaging, including the remaining guidance recommended in the Nanotechnology Task Force Report that has not yet been completed. FDA agrees that a critical element of its overall strategy for addressing the safety of nanomaterials used in food and food packaging is the likelihood and feasibility of GRAS determinations for such materials used in food and food packaging. FDA thus plans to issue a draft guidance that will help developers of food applications of nanotechnology determine the applicability of the GRAS concept.

The agency will continue to consider the viability of establishing an FDA-wide or even a foods definition of nanotechnology. While a single definition for nanomaterials for all FDA-regulated products may have certain utility, the key issue for safety evaluation is not whether a bright line on the nano-scale has been crossed but whether the modification at the nano-scale creates new properties that require testing to ensure safety under intended conditions of use.

End Notes

[1] Although we generally refer to GRAS substances in this report, under the Federal Food, Drug, and Cosmetic Act, it is the substance under conditions of its intended use, rather than the substance itself, that is eligible for the GRAS exemption.

[2] U.S. Department of Health and Human Services and U.S. Department of Agriculture, *Dietary Guidelines for Americans, 2005*, 6th ed. (Washington, D.C., U.S. Government Printing Office, 2005).

[3] Institute of Medicine of the National Academies, *Dietary Reference Intakes for Water, Potassium, Sodium, Chloride, and Sulfate* (Washington, D.C., The National Academies Press, 2004); Institute of Medicine of the National Academies, *Dietary Reference Intakes for Energy, Carbohydrate, Fiber, Fat, Fatty Acids, Cholesterol, Protein, and Amino Acids (Macronutrients)* (Washington, D.C., The National Academies Press, 2005).

[4] Some definitions have an upper limit higher than 100 nanometers.

[5] For more information on stevia and dietary supplements, see GAO, *Dietary Supplements: FDA Should Take Further Actions to Improve Oversight and Consumer Understanding*, GAO-09-250 (Washington, D.C.: Jan. 29, 2009).

[6] Department of Health and Human Services, FDA, *Food Protection Plan* (Rockville, Md., 2007). For more information on the *Food Protection Plan*, see GAO, *Federal Oversight of Food Safety: FDA Has Provided Few Details on the Resources and Strategies Needed to Implement its Food Protection Plan*, GAO-08-909T (Washington, D.C.: June 12, 2008); and *Federal Oversight of Food Safety: FDA's Food Protection Plan Proposes Positive First Steps, but Capacity to Carry Them Out Is Critical*, GAO-08-435T (Washington, D.C.: Jan. 29, 2008).

[7] Interagency Working Group on Import Safety, *Action Plan for Import Safety* (Washington, D.C., 2007).

[8] Guidance includes FDA, *Guidance for the Public, FDA Advisory Committee Members, and FDA Staff on Procedures for Determining Conflict of Interest and Eligibility for Participation in FDA Advisory Committees* (Rockville, Md., 2008) and FDA, *Guidance for the Public, FDA Advisory Committee Members, and FDA Staff: Public Availability of Advisory Committee Members' Financial Interest Information and Waivers* (2008).

[9] FDA's Priority-Based Assessment of Food Additives database contains administrative, chemical, and toxicological information on over 2,000 substances directly added to food.

[10] FDA has also required labeling of the trans fat content of foods since January 1, 2006.

[11] Besides the GRAS notification program, other regulatory tracks include pursuing a food contact substance notification or a food additive petition.

[12] U.S. Department of Health and Human Services and U.S. Department of Agriculture, *Dietary Guidelines for Americans, 2005*, 6th ed. (Washington, D.C., U.S. Government Printing Office, 2005).

[13] Institute of Medicine of the National Academies, *Dietary Reference Intakes for Water, Potassium, Sodium, Chloride, and Sulfate* (Washington, D.C., The National Academies Press, 2004); Institute of Medicine of the National Academies, *Dietary Reference Intakes for Energy, Carbohydrate, Fiber, Fat, Fatty Acids, Cholesterol, Protein, and Amino Acids (Macronutrients)* (Washington, D.C., The National Academies Press, 2005).

[14] The guidelines found that a potassium-rich diet blunts the effects of salt on blood pressure.

[15] Adequate intake represents the recommended average daily intake level based on observed or experimentally determined approximations or estimates of nutrient intake by a group (or groups) of apparently healthy people that are assumed to be adequate; adequate intake is used when a recommended daily allowance cannot be determined. Tolerable upper intake level is the highest average daily nutrient intake level that is likely to pose no risk of adverse health effects to almost all individuals in the general population; as intake increases above this level, the potential risk of adverse effects may increase.

[16] Department of Health and Human Services, FDA, *Food Protection Plan* (Washington, D.C., 2007).

[17] Interagency Working Group on Import Safety, *Action Plan for Import Safety* (Nov. 2007), http://www.importsafety.gov/ (accessed Sept. 18, 2009).

[18] FDA contracted for an independent review, by contemporary standards, of the available safety information related to substances it considered GRAS after new scientific information in the late 1960s raised questions about the safety of cyclamate salts, a class of sweeteners that FDA previously considered GRAS.

[19] FDA, *Nanotechnology: A Report of the U.S. Food and Drug Administration Nanotechnology Task Force* (Washington D.C., 2007).

[20] Council of Canadian Academies, *Small Is Different: A Science Perspective on the Regulatory Challenges of the Nanoscale* (Ottawa, Canada, 2008).

[21] European Food Safety Authority, "Scientific Opinion of the Scientific Committee on a Request from the European Commission on the Potential Risks Arising from Nanoscience and Nanotechnologies on Food and Food Safety," *The EFSA Journal*, no. 958 (2009).

In: The FDA and Generally Recognized as Safe … ISBN: 978-1-61209-924-8
Editor: James C. Monahan, pp. 69 - 89

Chapter 2

FEDERAL REGULATION OF SUBSTANCES GENERALLY RECOGNIZED AS SAFE (GRAS) AND THE USE OF CARBON MONOXIDE IN PACKAGING FOR MEAT AND FISH*

Vanessa K. Burrows and Cynthia Brougher

SUMMARY

The use of carbon monoxide (CO) in the packaging of meat and fish has generated considerable debate. The presence of CO results in the meat turning a bright red color that lasts longer than the color in untreated meat. Additionally, fish treated with CO gain a fresher appearance and a red tint. The meat industry, consumer groups, scientists, and policy makers disagree as to whether the use of CO in meat and fish packaging should be regulated by the Food and Drug Administration (FDA) and the United States Department of Agriculture (USDA), through labeling or otherwise, and whether CO should be a substance Generally Recognized As Safe (GRAS) under current and proposed FDA rules.

Two bills have been introduced in the 110th Congress regarding the use of carbon monoxide in meat, poultry products, and seafood: H.R. 3115 and H.R. 3610. The discussion draft of the Food and Drug Administration Globalization

* This is an edited, reformatted and augmented edition of a United States Congressional Research Service publication, Report RL34247, dated September 25, 2008.

Act of 2008, issued by Representatives Dingell, Pallone, and Stupak, similarly addresses the issue. The bills and the discussion draft propose to amend section 201 of the Federal Food, Drug, and Cosmetic Act (FFDCA). Under the proposals, if CO is used to treat meat, poultry, or seafood that is intended for human consumption, and if the conditions of that use would affect the color of the products, CO must be treated as a color additive under FFDCA, unless the product's label includes a statement that is prominently and conspicuously placed to notify the consumer of the use of CO and to warn the consumer of proper factors to judge the safety of the product. The bills and the discussion draft would allow the Secretary of Health and Human Services (HHS) to establish alternative labeling requirements five years after the effective date of the labeling requirement, if the Secretary finds that the labeling requirement is no longer necessary to prevent consumer deception. The discussion draft contains an additional provision related to GRAS determinations that would require the Secretary to publish, in the *Federal Register*, notice of receipt of a request for a substance to be determined by the Secretary to be GRAS. The Secretary would then have 90 days after publication of the notice to determine whether the substance is GRAS; the Secretary's determination would also be published in the *Federal Register*. Other bills also address GRAS substances: H.R. 2633, H.R. 3290, H.R. 3580, H.R. 6635, and S. 1342.

This report provides an overview of the FDA's regulation of GRAS substances, which are exempt from the premarket approval process for food additives. The report next discusses the FDA's 1997 proposed rule, which would create a notification procedure for GRAS substances through which manufacturers can notify the FDA of their "determination that a particular use of a substance is GRAS." The FDA has been using this GRAS notification procedure since the publication of the proposed rule on an "interim policy" basis. The roles of the USDA and FDA are also discussed, including the 2000 Memorandum of Understanding regarding review of substances used in the production of meat and poultry products. Finally, the report examines GRAS notices regarding intended uses of carbon monoxide.

INTRODUCTION

The use of carbon monoxide (CO) in the packaging of meat and fish has generated considerable debate. Carbon monoxide, in combination with nitrogen and carbon dioxide, is used in a packaging process for fresh meat called Modified Atmosphere Packaging (MAP).[1] In the MAP process, the meat is placed in a container with an "impermeable film similar to a vacuum package but ... the air [is evacuated] from the package and replac[ed] ... with a specified mixture of gases that provides for better control of product properties."[2] The presence of CO results in the meat turning a bright red color that lasts longer than the color in untreated meat. Additionally, fish treated with CO (for example, as part of a gas mixture called "tasteless smoke")[3] gain a fresher appearance and a red tint.[4] Conflicting studies have shown that consumers rely primarily on the appearance, including the red color of meat or fish, when choosing which package to purchase,[5] and alternatively, that consumers rely mostly on "sell by" dates.[6] The meat industry, consumer groups, scientists, and policy makers disagree as to whether the use of CO in meat and fish packaging should be regulated by the Food and Drug Administration (FDA) and the United States Department of Agriculture (USDA), through labeling or otherwise, and whether CO should be a substance Generally Recognized As Safe (GRAS) under current and proposed FDA rules.

The meat industry, some scientists, and other supporters argue that MAP reduces shrinkage of the meat, allows for a longer shelf life, "keep[s] meat fresh, protect[s] meat, [and] prevent[s] cross-contamination" because MAP packages are tamper resistant and leak-proof.[7] One scientist believes that MAP offers "better flavor, greater tenderness, and suppression of bacterial growth."[8] Supporters of MAP also assert that such products are more sustainable, less wasteful, and more flexible in terms of distribution because more packages can be transported per truck.[9] Additionally, they note that consumers prefer the bright red color of meat achieved in MAP.[10] Finally, MAP system supporters dispute the scientific basis for claims that the use of carbon monoxide is misleading or dangerous and declare the consumers use "sell by" dates when determining the freshness of many products.[11]

Opponents allege that the use of CO misleads consumers into thinking meat and fish are fresher than they are; that certain populations, such as those with a reduced sense of smell, will be at increased risk if they consume spoiled meat or fish that still appears fresh due to the use of CO; that consumers may eat undercooked meat because meat packed in MAP systems may brown faster when cooked than untreated meat; that "sell by" dates are not adequate to assist consumers in determining freshness; that consumers will be exposed to CO; that such MAP products are misbranded and adulterated under the Federal Food, Drug, and Cosmetic Act; and that the FDA is violating its own regulations on CO.[12] Another concern of consumer groups and some scientists is that CO provides a cover for spoiled or "temperature abused" meat and fish, meaning that the use of CO conceals visual cues of decomposition caused in part by exposure to changes in temperature or storage or transport at improper temperatures. Fish, such as tuna, may develop toxic levels of scombrotoxin (histamine) through time and/or temperature abuse, which can make consumers ill.[13] Opponents of the use of CO on meat and fish note that the European Union, Canada, Singapore, and Japan have prohibited or decided not to recognize or approve CO for use in fresh meat or fresh fish packaging.[14] Additionally, certain grocery store chains — including Giant, Safeway, Kroeger, and Publix — either do not sell or have announced that they will no longer sell MAP products.[15] Others have taken different steps. At a March hearing, a Target Corporation executive testified that its primary meat supplier had received approval from FSIS to add a label to its packaging that would state: "Color is not an accurate indicator of freshness. Refer to Use or Freeze By [date]."[16]

Legal Regulation of Food Additives and GRAS Substances

Both the FDA and USDA play a role in food safety and the types of substances that can be added to food. This section will focus on the FDA's regulation of food additives and GRAS substances, which the agency is responsible for under the Federal Food, Drug, and Cosmetic Act (FFDCA) and parts of Title 21, Code of Federal Regulations. FFDCA § 201(s) defines a food additive as:

> any substance the intended use of which results or may reasonably be expected to result, directly or indirectly, in its becoming a component or otherwise affecting the characteristics of any food (including any substance intended for use in producing, manufacturing, packing, processing, preparing, treating, packaging, transporting, or holding food; and including any source of radiation intended for such use)....[17]

The latter half of the above definition includes "food contact substances," which the FFDCA defines as "any substance intended for use as a component of materials used in manufacturing, packing, packaging, transporting, or holding food if such use is not intended to have any technical effect in such food."[18]

The definition of food additive excludes certain classes of substances: (1) pesticide chemical residues in or on a raw agricultural commodity or processed food, (2) pesticide chemicals, (3) color additives, (4) substances used in accordance with their sanction or approval under FDA and USDA laws prior to 1958, (5) new animal drugs, (6) dietary ingredients in dietary supplements, and

(7) substances GRAS under the conditions of the substances' intended use. These seven categories of substances are exemptions to FFDCA § 201(s) and do not have to obtain FDA approval as food additives before they can enter the market.[19] If a food additive does not meet one of the exemptions under the FFDCA, a rule must be in place that details the circumstances under which the food additive can be safely used.[20]

GRAS substances must be "generally recognized, among experts qualified by scientific training and experience to evaluate [their] safety."[21] FDA regulations recognize the difficulty of establishing the harmlessness of a substance and therefore define safety as "a reasonable certainty in the minds of competent scientists that the substance is not harmful under the intended conditions of use."[22] The person seeking GRAS status for a substance has the burden of proving the substance is GRAS under conditions of the substances' use.[23] A determination that a substance has GRAS status is not limited to FDA scientists. Experts may base their view of a general recognition of safety on either (1) scientific procedures or (2) common use of a substance in food prior to January 1, 1958.

The first type of GRAS substances is those that have "been adequately shown through scientific procedures ... to be safe under the conditions of [their] intended use."[24] Scientific procedures include published and unpublished human, animal, analytical, and other scientific studies that are "appropriate to establish the safety of a substance."[25] A GRAS determination based on scientific procedures "require[s] the same quantity and quality of scientific evidence as is required to obtain approval of a food additive regulation for the ingredient."[26] The GRAS determination must "ordinarily" be based on published studies, but can be corroborated by unpublished studies and other information.[27] FDA regulations do not require a unanimous opinion from the scientific community that a substance is GRAS under the conditions of its intended use; rather, the person seeking GRAS status "must show that there is a consensus of expert opinion regarding the safety of the use of the substance."[28] However, "a severe conflict among experts regarding the safety of the use of a substance, precludes a finding" that a substance is GRAS.[29]

The second type of GRAS substances is those that were "used in food prior to January 1, 1958, [and shown] through either scientific procedures or experience based on common use in food[] to be safe under the conditions of [their] intended use."[30] FDA regulations define the phrase "common use in food" as "a substantial history of consumption of a substance for food use by a significant number of consumers."[31] In this instance, a GRAS determination ordinarily turns on "generally available data and information."[32] These substances are known as prior-sanctioned substances. They can include substances used in food where the use prior to January 1, 1958, "occurred exclusively or primarily outside of the United States if the information about the experience establishes that the use of the substance is safe."[33] Published information regarding substances used outside the United States must be corroborated.[34]

The FDA lists some GRAS substances in 21 C.F.R. Part 182. However, this list of GRAS substances is not exhaustive as "[i]t is impracticable to list all substances that are [GRAS] for their intended use."[35] The list of GRAS substances in 21 C.F.R. Part 182 includes spices, essential oils, natural extracts, synthetic flavoring substances, substances that migrate from dry food packaging and paper products, multipurpose substances, anticaking agents, chemical preservatives, emulsifying agents, stabilizers, sequestrants, and nutrients.[36]

The FDA Commissioner can affirm the GRAS status of a substance based on a petition from a manufacturer or others or on his or her own initiative.[37] Substances affirmed as GRAS, listed in 21 C.F.R. Part 184, differ from the GRAS substances listed in Part 182 because their GRAS status has been sustained through a notice-and-comment rulemaking. The concept of affirming the GRAS status of substances began in 1969, when questions arose about whether cyclamate salts, a substance that had been considered GRAS, were safe because "they were implicated in the formation of bladder tumors in rats."[38]

The affirmation of GRAS status occurs through the notice and comment rulemaking process, in which the Commissioner publishes a notice of the substance proposed to be affirmed as GRAS in the *Federal Register*, allows 60 days for comments, evaluates the comments (and the petition, if one was filed), and either (1) publishes a notice in the *Federal Register* affirming the substance is GRAS if there is "convincing evidence" or (2) "concludes that there is a lack of convincing evidence that the substance is GRAS and that it should be considered a food additive" subject to premarket approval by the FDA under FFDCA § 409.[39] If the agency affirms that the use of a substance is GRAS, the substance is added to a list in the *Code of Federal Regulations* as a substance affirmed as GRAS "for the purposes and under the conditions prescribed,"[40] allowing for the possibility that use of a substance under a condition other than the one specified in the regulation may not be GRAS.[41] The FDA has reviewed the direct food substances on the list in Part 184 and determined that they are GRAS "for the purposes and under the conditions prescribed."[42] These ingredients are also GRAS as indirect food ingredients, also known as food contact substances, within certain limitations.[43] Part 186 of Title 21, Code of Federal Regulations, lists the indirect food substances/food contact substances affirmed as GRAS, such as wrappers, containers, and other food-contact surfaces.[44]

If the Commissioner reviews a food ingredient and finds that it is a GRAS substance, under 21 C.F.R. § 184.1, the final rule approving the GRAS substance for the purposes and under the conditions prescribed may contain limits on the application and use of the substance. First, the regulation identifies the characteristics of the ingredient in such a way that it can be differentiated from other versions of the ingredient that the FDA has not affirmed as GRAS.[45] Second, the substance affirmed as GRAS "must be used in accordance with current good manufacturing practices."[46] Third, a FDA regulation affirming GRAS status "when the safety of an ingredient has been evaluated on the basis of limited conditions of use" will specify the limited conditions of use. Use of the ingredient under a condition other than the one specified in the regulation may not be GRAS.[47] In such a case, the manufacturer must "independently establish that that use is GRAS or shall use the ingredient in accordance with a food additive regulation."[48] Fourth, the substance affirmed as GRAS for the purposes and conditions prescribed cannot be used "in a manner that may lead to deception of the consumer" or FFDCA violations.[49] Finally, ingredients listed as GRAS cannot be combined, in order to achieve the same technological effect in a food, at levels greater than were permitted for a single ingredient.[50]

THE FDA'S 1997 PROPOSED RULE

The procedure outlined in a FDA proposed rule from 1997 would eliminate the notice and comment rulemaking process described above for substances affirmed as GRAS.[51] The proposed rule would also end the GRAS petition process and create a new GRAS notification procedure.[52] Although the notice and comment rulemaking process for GRAS substances is still in effect in the FDA regulations, the FDA has effectively been using the GRAS notification procedure outlined in the proposed rule since 1998 without issuing a final rule. Since the FDA has not issued a final rule, it is important to note that the FDA's procedure set forth in the 1997 proposed rule is only guidance and not law. The agency has also issued guidance for industry in the form of frequently asked questions about GRAS that includes a discussion of the GRAS notification program.[53] More than 250 GRAS notifications have been submitted under the procedure outlined in the 1997 proposed rule. The FDA has issued one of the three responses described below for most of these notices, and both a numerical and alphabetical list of notices received and agency responses can be found on the FDA's website.[54]

Under the notification procedure in the proposed rule, industry submits a GRAS notification to the FDA that states the company's view that the substance is GRAS. These notifications identify the notifier and describe the substance that is the subject of the notice, the applicable conditions of use, and the basis for the GRAS determination, including a summary of supporting information "that forms the basis for an exemption from a statutory requirement."[55] The notifier "explicitly accepts responsibility for the GRAS determination," unlike the protocol in the current regulations, in which such responsibility falls on the agency because an interested person has petitioned the FDA to affirm a use of a substance as GRAS or the FDA itself has affirmed a substance's use as GRAS.[56]

Rather than requiring that the FDA affirm that a substance is GRAS through a notice-and-comment rulemaking, the 1997 proposed rule provides that the FDA does not make a finding that a substance in a GRAS notification made under the proposed rule process actually is a GRAS substance. Instead, the agency states that (1) it has "no questions" about the notifier's conclusion that a substance is GRAS, (2) the notice does not provide a basis for a GRAS status determination, or (3) the notifier has stopped the GRAS notification process.[57] If the agency's review of a GRAS notification does not furnish appropriate information to find a basis for a GRAS determination, it will issue such a response, potentially in light of the following reasons to question the use of the substance:

> FDA may question the GRAS status of use of a substance if the information provided in a notice: (1) Does not adequately establish technical evidence of safety; (2) is not generally available; (3) does not convince the agency that there is the requisite expert consensus about the safety of the substance for its intended use; or (4) is so poorly presented that the basis for the GRAS determination is not clear. FDA also may be aware of information that is not included in the notice but raises important public health issues that lead the agency to question GRAS status of use of the substance.[58]

The FDA notes that notifiers "receive as a benefit a response that documents the agency's awareness of the [GRAS] determination" by the notifier.[59]

If, as in the majority of the FDA's responses to GRAS notification submissions, the FDA has no questions about the notification, this determination does not mean that the FDA has approved the substance in the notification as GRAS.[60] In other words, none of the uses of the substances reviewed by the FDA through a GRAS notification are deemed to actually be GRAS by the FDA.[61] Moreover, in contrast to the FDA's GRAS affirmation regulations, which allow the FDA to place potential limits on the use of a GRAS substance, the GRAS notification procedures in the FDA's proposed rule do not appear to allow this, as the FDA only responds in one of three ways noted above. Nonetheless, an FDA response of "no questions" could give a substance an imprimatur of safety from the federal government. Such a response may also give manufacturers confidence that the substance is acceptable, and they would be able to tell their suppliers and others of the FDA's response to the notification. Additionally, an FDA response of "no questions" may convey to manufacturers a feeling of less uncertainty and less potential liability about using such a GRAS substance that has been through the GRAS notification process, as the agency may not be as likely to seize a substance or find a product adulterated or misbranded if the FDA itself has said it has "no questions."[62]

As mentioned above, the agency has yet to issue a final rule on the notification procedure; however, the FDA has "invite[d] interested persons" to submit such notifications as described in the proposed rule on an "interim policy" basis until the publication of the final rule.[63] The agency has accepted more than 250 notification submissions under the proposed rule procedures.[64] In its proposal, the FDA has stated that it "will determine whether its experience in administering such notices suggests modifications to the proposed procedure."[65] The agency's description and adoption of the new GRAS notification process (as delineated in the proposed rule) on an interim policy basis may be characterized as the equivalent of a guidance document.[66]

Statistics on FDA GRAS Notices

The chart below provides the number of FDA response letters in each of the three categories discussed above, as well as a fourth category for the number of GRAS notices that are awaiting a response from the FDA, and the percent of the total number of letters issued by the FDA under its procedure in the 1997 proposed rule. One GRAS notification, GRN No. 13, was counted twice — once in the "FDA has no questions" category and once in the "Notice does not provide a basis for a GRAS determination" category — because the FDA had no questions for three botanical substances in the notice (Chrysanthemum, Licorice, and Jellywort) but the FDA stated that the notice did not provide a basis for a GRAS determination for six other substances (Honeysuckle; Lophatherum; Mulberry leaf; Frangipani; Selfheal; Sophora flower bud).

The FDA's response to GRAS notifications that were initially submitted, but then were either withdrawn or determined not to provide a basis for a GRAS determination, were only included for the resubmitted notices for the same substances. For example, Hawaii International Seafood, Inc. initially submitted its GRAS notification for tasteless smoke as GRAS Notice No. 5, but then at the company's request, the FDA ceased to evaluate the notice. Hawaii International Seafood, Inc. then resubmitted its GRAS notification for tasteless smoke as GRAS Notice No. 15, and the FDA had no questions. Only the FDA's response to the resubmitted notification is included on the chart below. There were 18 instances of GRAS notifications being resubmitted, which explains the difference in the chart's total number of notices (238) and the number of GRAS notifications listed on the FDA website (256).

Categories of FDA Responses or Response is Pending[a]	FDA has no questions[b]	Notice does not provide a basis for a GRAS determination	At notifier's request, FDA ceased to evaluate the notice	Pending	Total
Number of FDA Lett-ers in Each Category[a]	179	10	22	27	238
Percentage of Total Lettersc	75.21	4.20	9.24	11.34	99.99[c]

a. The categories of FDA letters and the number of FDA letters in each category were obtained from the FDA Center for Food Safety and Applied Nutrition's Numerical Listing of GRAS Notices for July 2008. [http://www.cfsan.fda.gov/~rdb/opa-gras.html].

b. This category includes notices in which the FDA had no questions but stated that some uses of a GRAS substance may require a color additive listing.[67]

c. Percentages were calculated by CRS and rounded to two decimal points.

THE ROLE OF USDA IN FOOD ADDITIVE SAFETY DETERMINATIONS

Under the current legislative and regulatory schemes, the FDA shares responsibility for some food safety issues with the United States Department of Agriculture (USDA). While the FDA is responsible for safety of the vast majority of food categories, the USDA is specifically authorized to regulate the safety and wholesomeness of meat and poultry products that are intended for use as human food.[68] Under this authority, the USDA, and consequently the Food Safety and Inspection Service (FSIS),[69] is required to provide a mark of inspection on meat and poultry products.[70] The mark of inspection reflects a determination that the product is not adulterated[71] or misbranded.[72]

Dual Process of Review for Meat and Poultry Products

There is a two-step process for approving the use of additive substances in meat and poultry products: (1) FDA determines the safety of substances and prescribes safe conditions of use, and (2) FSIS determines whether new substances or new applications of substances are suitable for use in meat and poultry products.[73] In other words, FDA makes determinations based on the safety of the substance itself, while FSIS approves the substance's application to the meat or poultry product.

In 2000, the roles of FDA and FSIS in this joint review process of substances used in meat and poultry products were laid out in a Memorandum of Understanding (MOU).[74] The MOU provides for standard operating procedures regarding submissions to FDA or FSIS that, for example, petition for the approval of food and color additives intended for use in meat or poultry products, as well as GRAS notifications "regarding the use of a substance in the production of meat or poultry products."[75] The MOU generally instructs the agency that receives a request for review of a substance used in meat or poultry products to seek review by the other agency regarding the substance as well. For example, when FSIS receives a request for an acceptability determination regarding the application of a substance in the production of meat or poultry products, it confirms the status of the substance's safety with FDA.[76] Conversely, if FDA receives a request for a suitability determination regarding the use of a substance in meat or poultry products, the request must be transferred to FSIS.[77]

The MOU provides that when FDA receives a GRAS Notice regarding the use of a substance in the production of meat or poultry products, FDA and FSIS proceed jointly, as they would regarding requests for approval of a food or color additive intended for use in the production of meat or poultry products. FDA informs and consults with FSIS, and FSIS provides written comments to FDA within 60 days. FDA's response to the notifier includes information regarding the notifier's responsibilities under the Federal Meat Inspection Act and Poultry Products Inspection Act and "may include concerns about the suitability of the use of the substance in the production of meat or poultry products and, when applicable, any restrictions or conditions of use in the production of meat or poultry products that FSIS recommends in writing."[78]

FSIS Review of Substances in Meat or Poultry Products

Under the dual review process, if FDA approves a substance, such as a food or color additive, or lists the substance as GRAS for use in food, the substance is not automatically acceptable for use in meat and poultry products. If FDA's approval of a food or color additive, or if the FDA's GRAS listing does not specifically mention meat or poultry products, FSIS needs an affirmative written statement from FDA that it did consider the substance's use in meat or poultry or that it has no objections with regard to safety when the substance is used in meat or poultry.[79] FSIS then needs to determine suitability and whether rulemaking is required.[80] Whether a substance is suitable depends on "the effectiveness of the substance in performing the intended technical purpose of use, at the lowest level necessary, and the assurance that the conditions of use will not result in an adulterated product or one that misleads customers."[81]

To satisfy the requirement of suitability, FSIS needs certain data as evidence that the substance or use of the substance is suitable for its intended technical purposes.[82] The data must show the effectiveness of the substance in achieving the intended purpose of its use.[83] The data must show that the use is at the lowest level necessary to achieve the intended effect under the proposed conditions of use.[84] The data must show that the use cannot result in adulteration or misbranding.[85] FSIS regulations currently prohibit the use of substances that conceal damage or inferiority or make a product appear better or of greater value than it is.[86] The regulations also provide that substances that are intended to be used to impart color in any meat or poultry product cannot be used unless approved as a color additive (under FDA regulations) or approved by FSIS regulations.[87] This data must be provided for each separate product in which the use of the substance is intended.[88] Based on the merits of these data, FSIS can permit the use of the substance or the new use of a substance under the proposed conditions of use and in conformance with standards and labeling requirements.

With respect to whether rulemaking is required, if FDA has found or confirmed the safety of the substance, FSIS regulations are not amended.[89] If rulemaking is not required, FSIS notifies the requestor in writing of its determination in what is known as an acceptability determination.[90] If the use of the substance is prohibited or limited or if the substance is not normally found in the product, FSIS regulations may be necessary.[91] If rulemaking is required, the substance is added to the current list of approved substances after the formal rulemaking process is completed.[92] Because not all approved substances are listed in the published regulations under this process, FSIS maintains a directive system of all approved substances that are accepted as safe and suitable by FSIS on its website.[93]

GRAS Substances

As discussed above, a determination that a substance is GRAS may be made by the FDA, through the affirmation of the GRAS status of a substance, or by industry (including via a GRAS notification), based on scientific procedures or common use of a substance in food prior to January 1, 1958.[94] FSIS cannot rely on the industry's determination of a substance as GRAS because of statutory requirements requiring USDA inspection of meat and poultry products.[95] Meat and poultry products are required to have a mark of inspection that "reflects a determination by FSIS that the food product is not adulterated, and thus that all substances used to make the product are safe and suitable."[96] As a result, "F SIS must have from FDA, at the very least, a written statement of no objection with regard to the safety of the use of the substance."[97]

GRAS Notices Regarding Intended Uses of Carbon Monoxide

Under the process outlined in the FDA's 1997 proposed rule, manufacturers have submitted GRAS notifications to the FDA that state their view that carbon monoxide is a GRAS substance. The FDA has responded that it has "no questions" about the conclusion that CO is GRAS. The FDA's responses to the GRAS notifications informed the industry that it had the continuing responsibility to ensure the substance's safety and compliance with other legal and regulatory requirements.

The FDA first determined that it had no questions regarding a GRAS notification for the use of carbon monoxide in March 2000. The notifier, Hawaii International Seafood, Inc., stated its determination that the use of "tasteless smoke" (of which carbon monoxide is a component) on raw seafood is GRAS.[98] The company defined its intended use as involving a procedure before the fish is frozen that would preserve the color, taste, aroma, and texture of raw seafood.[99] In addition to determining that it had no questions, the FDA stated that the company's use of tasteless smoke constituted a preservative and noted that the fish must be labeled so that it complies with misbranding provisions of the FFDCA and the FDA's labeling regulations.

The FDA next determined that it had no questions regarding the use of carbon monoxide as a GRAS substance in meat packaging in a letter to Pactiv Corporation in February 2002.[100] The agency also stated that it had not made an independent determination of the GRAS status of the use of CO described in the notification.[101] FDA noted the industry's conclusion that the use of carbon monoxide allows meat to maintain a desirable red color during storage but once the product was removed from storage, the color of the meat "deteriorates at a similar rate to that of meat that has not been exposed to CO."[102] FSIS concluded that the use of carbon monoxide in the MAP system as it had been described by Pactiv in its GRAS notification "would be acceptable for packaging red meat cuts and ground meat."[103] FSIS agreed with the company that "there is no lasting functional effect in the food and there is an insignificant amount of carbon monoxide present in the finished product under the proposed conditions of use."[104] FDA restated that it had no questions regarding the industry's determination that carbon monoxide is GRAS in July 2004 and September 2005 in response letters to Precept Foods, LLC, and Tyson Foods, Inc., respectively.[105] Currently, two additional GRAS notifications regarding carbon monoxide are pending.[106]

Although notifiers seeking a response from the FDA on GRAS notices for CO have submitted notices describing other conditions of use of CO, it appears possible that a manufacturer could potentially rely on a FDA response that the agency "has no questions" to use a GRAS substance in a manner other than the use described in the GRAS notice for which the FDA had no questions. For example, two of the CO GRAS notices, 83 and 143, discuss a level of CO that is 0.4 percent in a MAP system. Conceivably, a company could interpret the agency's lack of questions regarding the 0.4 CO level and use a CO level of 0.45 percent in a MAP system. However, if the FDA made a determination that the use of 0.45 percent CO, or even 0.4 percent CO, violated the FFDCA, the agency could attempt to seek criminal and civil penalties for violations such as adulteration and misbranding. The FFDCA also provides the FDA with other enforcement mechanisms such as seizure and injunctions.

PROPOSED LEGISLATION IN THE 110TH CONGRESS

Two bills have been introduced in the House of Representatives regarding the use of carbon monoxide in meat and poultry products: H.R. 3115 (the Carbon Monoxide Treated Meat, Poultry, and Seafood Safe Handling, Labeling, and Consumer Protection Act) and H.R. 3610 (the Food and Drug Import Safety Act of 2007). Additionally, the discussion draft of the Food and Drug Administration Globalization Act of 2008, issued by Representatives Dingell, Pallone, and Stupak, similarly addresses the issue. Other bills also address GRAS substances: H.R. 2633, H.R. 3290, H.R. 3580, H.R. 6635, and S. 1342.

H.R. 3115, H.R. 3610, and the discussion draft propose to amend FFDCA § 201.[107] Under the proposals, if carbon monoxide is used to treat meat, poultry or seafood that is intended for human consumption and if the conditions of that use would affect the color of the products, carbon monoxide must be treated as a color additive under FFDCA,[108] unless the product's label includes a statement that is "prominently and conspicuously" placed to notify the consumer of the use of carbon monoxide and to warn the consumer of proper factors to judge the safety of the product.[109] The bills and the discussion draft would allow the Secretary of Health and Human Services (HHS) to establish alternative labeling requirements five years after the effective date of the labeling requirement, if the Secretary finds that the labeling requirement is no longer necessary to prevent consumer deception.

The discussion draft contains an additional provision related to GRAS determinations that would require the Secretary to publish, in the *Federal Register*, notice of receipt of a request for a substance to be determined by the Secretary to be GRAS. The Secretary would then have 90 days after publication of the notice to determine whether the substance is GRAS; the Secretary's determination would also be published in the *Federal Register*. It is unclear if the discussion draft is referring to a petition for affirmation of GRAS status under 21 C.F.R. § 170.35 as the "request for a substance to be determined by the Secretary to be a GRAS substance," or an alternate situation. (See page 5 in the PDF version of this report.) If the FDA Commissioner receives a petition to affirm the GRAS status of a substance "that directly or indirectly become[s] [a] component[] of food," the Commissioner must publish a notice of the filing of the petition in the *Federal Register* within 30 days after the date of filing of the petition.[110] There is a 60-day comment period after the notice of filing in the current regulations.[111] The current regulations state that the FDA Commissioner will publish an order listing the substance as GRAS if the petition and all available information "provide[s] convincing evidence that the substance is GRAS."[112] Alternatively, if the Commissioner "concludes that there is a lack of convincing evidence that the substance is GRAS and that it should be considered a food additive subject to" FFDCA § 409, the Commissioner must publish a notice of this determination in the *Federal Register*.[113]

End Notes

[1] Agency Response Letter GRAS Notice No. GRN 000083 from Alan M. Rulis, Director, Office of Food Additive Safety, Center for Food Safety and Applied Nutrition, FDA, to Eric Greenberg, Ungaretti and Harris (on behalf of Pactiv Corp.) (Feb. 12, 2002), [http://www.cfsan.fda.gov/~rdb/opa-g083 .html], [hereinafter GRAS Notice No. 83].

[2] To Review Technologies in the Meat Industries: Hearing before the H. Comm. on Agriculture, 110th Cong. 2 (2007) (statement of Joe Sebranek, Dep't of Animal Science and Dep't of Food Science, Iowa State Univ.), [http://agriculture 10/ h7 1 030/Sebranek.doc].

[3] Agency Response Letter GRAS Notice No. GRN 000015 from Janice F. Oliver, Deputy Director, Center for Food Safety and Applied Nutrition, FDA, to Martin J. Hahn, Hogan & Hartson LLP (on behalf of Hawaii International Seafood, Inc.) (Mar. 10, 2000), [http://www.cfsan.fda.gov/~rdb/opa-g0 15 .html], [hereinafter GRAS Notice No. 15].

[4] Harold McGee, *The Red-Meat Miracle, and Other Tales from the Butcher Case*, N.Y. Times, Dining, 4 (Apr. 4, 2007).

[5] Press Release, Consumer Federation of America, Most Consumers Are Concerned About Practice of Adding Carbon Monoxide to Meat, New Survey Finds (Sept. 25, 2006), [http://www.consumerfed.org/pdfs/CO_Meat_Consumer_Press_Release_9.25 .06.pdf]. ("Sixty-three percent (63%) agreed with the statement that 'the freshness of meat is directly related to the color of the meat.'").

[6] To Review Technologies in the Meat Industries: Hearing before the H. Comm. on Agriculture, 110th Cong. 11, 21 (2007) (statement of Phil Minerich, Vice President, Research and Development, Hormel Foods Corp.), [http://agriculture 11 0/h7 1 030/Minerich.pdf] (A 2005 Food Marketing Institute study cited by Minerich in his testimony concluded that "8 1% of consumers rely on sell by dates.").

[7] GRAS Notice No. 83, *supra* note 1; Minerich, *supra* note 6, at 6.

[8] Sebranek, *supra* note 2, at 3.

[9] Minerich, *supra* note 6, at 17; GRAS Notice No. 83, *supra* note 1.

[10] *See* Minerich, *supra* note 6, at 30 (quoting Dr. Gary Acuff, Professor of Microbiology, Texas A&M University in a May 26, 2006, letter to the editor of *Meating Place* magazine); Sebranek, *supra* note 2, at 1.

[11] Sebranek, *supra* note 2, at 1; Minerich, *supra* note 6, at 25.

[12] Citizen Petition from Donald R. Berdahl, Executive Vice President, Kalsec, Inc., to Laura M. Tarantino, Director, Office of Food Safety, Center for Food Safety and Applied Nutrition, FDA (Nov. 20, 2006), [http://www.co-meat filing.PDF] at 5-6, 8; Sebranek, *supra* note 2, at 2-3; McGee, *supra* note 4.

[13] Center for Food Safety and Applied Nutrition, FDA, Fish and Fisheries Products Hazards and Controls Guidance, ch. 7, (June 2001), [http://www.cfsan.fda.gov/~comm/haccp4g. html].

[14] Carbon Monoxide in Fresh Meat, Selected Countries Prohibiting Carbon Monoxide (CO) Gas in Fresh Meat and Fresh Fish Packaging, [http://www.co-meat

[15] Julie Schmidt, *Carbon Monoxide Keeps Meat Red Longer; Is that Good?*, USA Today, Oct. 30, 2007.

[16] Regulatory Failure: Must America Live with Unsafe Food? Hearing before the H. Comm. on Energy and Commerce, Subcomm. on Oversight and Investigations, 110th Cong. (Mar. 12, 2008) (statement of Danielle Lachman, Divisional Merchandise Manager, Target Corporation), [http://energycommerce.house.gov/cmte_mtgs/1 1 0-oi-hrg.03 1208 .LachmanTestimony.pdf].

[17] FFDCA § 201(s); 21 U.S.C. § 321(s).

[18] FFDCA § 409(h)(6); 21 U.S.C. § 348(h)(6).

[19] FFDCA § 409; 21 U.S.C. § 348.

[20] FFDCA § 409(a)(2); 21 U.S.C. § 348(a)(2).

[21] FFDCA § 201(s); 21 U.S.C. § 321(s).

[22] 21 C.F.R. § 170.3(I). This reasonable certainty of no harm standard applies to food additives, color additives, and food contact substances, in addition to GRAS substances.

[23] *See* Substances Generally Recognized as Safe, 62 Fed. Reg. 18937, 18939 (proposed Apr. 17, 1997).

[24] FFDCA § 201(s); 21 U.S.C. § 321(s).

[25] 21 C.F.R. § 170.3(h).

[26] 21 C.F.R. § 170.30(b).

[27] 21 C.F.R. § 170.30(b).

[28] Substances Generally Recognized as Safe, 62 Fed. Reg. 18937, 18939 (proposed Apr. 17, 1997).

[29] *Id.*

[30] FFDCA § 201(s); 21 U.S.C. § 321(s).

[31] 21 C.F.R. § 170.3(f).

[32] 21 C.F.R. § 170.30(c)(1).

[33] 21 C.F.R. § 170.30(c)(2).

[34] 21 C.F.R. § 170.30(c)(2).

[35] 21 C.F.R. § 182.1(a).

[36] *See* 21 C.F.R. § 170.3 (providing definitions of several of the above terms). Sequestrants are "[s]ubstances which combine with polyvalent metal ions to form a soluble metal complex, to improve the quality and stability of products." 21 C.F.R. § 170.3(o)(26).

[37] 21 C.F.R. § 170.35(a). "The rulemaking process in § 170.35(c) whereby manufacturers may petition FDA to affirm that a substance is GRAS under certain conditions of use was designed as a voluntary administrative process whose purpose was to provide a mechanism for official recognition of lawfully made GRAS determinations." 62 Fed. Reg. 18941.

[38] Substances Generally Recognized as Safe, 62 Fed. Reg. 18937, 18939 (proposed Apr. 17, 1997). The FDA Commissioner "conclude[d] that cyclamates c[ould] no longer be regarded as generally recognized as safe for use in food" and amended the Code of Federal Regulations to delete such substances from the GRAS list. 34 Fed. Reg. 17063 (Oct. 21, 1969).

[39] 21 C.F.R. § 170.35.

[40] 21 C.F.R. § 170.35(b), (c); Substances Generally Recognized as Safe, 21 C.F.R. Part 182; Direct food substances affirmed as generally recognized as safe, 21 C.F.R. Part 184; Indirect food substances affirmed as generally recognized as safe, 21 C.F.R. Part 186.

[41] 21 C.F.R. § 184.1(b)(1); 21 C.F.R. § 170.30(I). In such a case, a manufacturer or other person must "independently establish that that use is GRAS or shall use the ingredient in accordance with a food additive regulation." 21 C.F.R. § 184.1(b)(1); *see also* 21 C.F.R. § 170.30(I).

[42] 21 C.F.R. § 184.1(a). Presently, the FDA is evaluating a Citizen's Petition with regard to the affirmed GRAS status of diacetyl, "a primary component of butter flavoring in a number of foods, including microwave popcorn," that has been linked to brochiolitis obliterans, "a rare, sometimes fatal respiratory disease." E-mail from FDA Office of Legislation, Feb. 8, 2008 (on file with author); Andrew Schneider, *Flavoring Additive Puts Professional Cooks at Risk*, Seattle Post-Intelligencer, (Dec. 21, 2007), [http://seattlepi. nwsource.com/national/ 344277_diacetyl21.html]. While 21 C.F.R. § 184.1278 does not place any limits on the use of diacetyl, and the FDA has stated it is "not aware of any evidence that consumption of diacetyl (as opposed to inhalation) is unsafe," the FDA would have the power to prescribe limits under 21 C.F.R. § 184 on the purposes and conditions for which diacetyl could be used — for example, not as a component of flavoring for microwave popcorn or butter substitutes that release a potentially harmful vapor from diacetyl when heated. Schneider, *supra*.

[43] *Id.*

[44] 21 C.F.R. § 186.1(b).

[45] 21 C.F.R. § 184.1(a).

[46] 21 C.F.R. § 184.1(b).

[47] 21 C.F.R. § 184.1(b)(1).

[48] 21 C.F.R. § 184.1(b)(1).

[49] 21 C.F.R. § 184.1(c).

[50] 21 C.F.R. § 184.1(d).

[51] Substances Generally Recognized as Safe, 62 Fed. Reg. 18937, 18939 (proposed Apr. 17, 1997); *see also* Agency Information Collection Activities; Proposed Collection; Comment request; Notice of a Claim for Generally Recognized as Safe Exemption Based on a Generally Recognized as Safe Determination, 70 Fed. Reg. 73009 (Dec. 8, 2005).

[52] The FDA's notification procedure outlined in the 1997 proposed rule is not unique. The Food and Drug Administration Modernization Act of 1997 (FDAMA) created a notification procedure for food contact substances, and developers of genetically engineered crops need to obtain authorization from USDA's Animal and Plant Health Inspection Service through either a permit or notification process.

[53] Center for Food Safety and Applied Nutrition, FDA, Guidance for Industry: Frequently Asked Questions About GRAS (Dec. 2004), [http://www.cfsan.fda.gov/~dms/grasguid. html].

[54] Center for Food Safety and Applied Nutrition, FDA, Numerical Listing of GRAS Notices (October 2007), [http://www.cfsan.fda.gov/~rdb/opa-gras.html]; Center for Food Safety and Applied Nutrition, FDA, Alphabetical Listing of GRAS Notices (Oct. 2007), [http://www.cfsan.fda.gov/~rdb/opagras1.html].

[55] 62 Fed. Reg. 18947. "[T]he notifier must consent to grant the FDA access to the data and information that are the basis of the GRAS determination," and the agency has stated that it "intends to conduct random audits of [this] data and information." *Id.*

[56] 62 Fed. Reg. 18946.

[57] Substances Generally Recognized as Safe, 62 Fed. Reg. 18937 (proposed Apr. 17, 1997).

[58] 62 Fed. Reg. 18950.

[59] 62 Fed. Reg. 18947.

[60] 62 Fed. Reg. 18951.

[61] The use of the term 'review' here does not mean that the agency will necessarily "conduct its own detailed evaluation" of, for example, raw data of toxicological studies or data used to support the notifier's GRAS determination. 62 Fed. Reg. 18948-49.

[62] *See* Community Nutrition Institute v. Young, 818 F.2d 943, 949 n.10 (D.C. Cir. 1987) (characterizing, in a case where the FDA set levels above which it could take action on adulterated corn, the U.S. Supreme Court's description of action levels as "agency assurance": "In setting an action level, the FDA essentially assures food producers that it ordinarily will not enforce the general adulteration provisions of the Act against them.")(internal citations omitted).

[63] 62 Fed. Reg. 18954-55.

[64] The FDA's acceptance of GRAS notifications since the publication of the 1997 proposed rule could be seen as an "experiment[] while the rulemaking is in progress" that could bolster the need for a subsequent notice-and-comment period before the agency publishes a final rule. JEFFREY S. LUBBERS, A GUIDE TO FEDERAL AGENCY RULEMAKING 292-93 (4th ed. 2006). Agencies frequently publish requests for additional rounds of notice-andcomment during the rulemaking process. The FDA's NPRM on GRAS substances asked for the submission of written comments by July 16, 1997. 62 Fed. Reg. 18938.

[65] 62 Fed. Reg. 18954.

[66] If the interim policy procedures are considered to be a guidance document, which is a type of general statement of policy under the Administrative Procedure Act (APA), 5 U.S.C. § 553(b), then the FDA arguably would not need to complete the rulemaking because the notice and comment provisions of the APA would not apply. *See* LUBBERS, *supra* note 64, at 94. Thus the agency would not appear to be violating the APA if the guidance document procedures are prospective and voluntary and if the interim policy preserves the FDA's discretion. Moreover, the GRAS affirmation regulations are still law and may still be used by interested persons. However, completing the rulemaking may clarify agency policy. Additionally, although the APA does not impose a limit on the time between an agency's publication of a proposed rule and its issuance of a final rule, the 1997 proposed rule may be considered to be stale, and a new rulemaking may be necessary. 5 U.S.C. § 555(b); LUBBERS, *supra* note 64, at 293, n. 82, 357-58.

[67] *See, e.g.*, Agency Response Letter GRAS Notice No. GRN 000156 from Laura M. Tarantino, Director, Office of Food Additive Safety, Center for Food Safety and Applied Nutrition, FDA, to George Burdock, Burdock Group (on behalf of LycoRed Natural Products Industries, Ltd.), Feb. 7, 2005, [http://www.cfsan.fda.gov/~rdb/opa-g1 56.html].

[68] This authority is provided for by the Federal Meat Inspection Act (21 U.S.C. § 601 et seq.) and the Poultry Products Inspection Act (21 U.S.C. § 451 et seq.), respectively.

[69] The Secretary of the USDA delegates this authority to FSIS under 9 C.F.R. § 300.2.

[70] *See* 21 U.S.C. § 606; 21 U.S.C. § 457.

[71] A product can be considered "adulterated" if it "bears or contains any poisonous or deleterious substance which may render it injurious to health;" contains any additives considered unsafe; "consists in whole or in part of any filthy, putrid, or decomposed substance or is for any other reason unsound, unhealthful, unwholesome, or otherwise unfit for human food;" or "has been prepared, packed, or held under insanitary conditions." *See* 21 U.S.C. § 601(m).

[72] A product can be considered "misbranded" if its label is false or misleading; contains an inaccurate description of the product; does not identify its manufacturer, packer or distributor and an accurate statement of quantity of the contents; or does not contain other information that may be required by the act. *See* 21 U.S.C. § 601(n).

[73] *See* 9 C.F.R. § 424.21.

[74] Memorandum of Understanding Between the Food Safety and Inspection Service United States Department of Agriculture and the Food and Drug Administration United States Department of Health and Human Services Regarding the Listing or Approval of Food Ingredients and Sources of Radiation Used in the Production of Meat and Poultry Products, [http://www.fsis.usda.gov/Regulations_&_Policies/Labeling_FDA

[75] *Id.*

[76] Standard Operating Procedures D, [http://www.fsis.usda.gov/Regulations_&_Policies/ Labeling_ FDA_MOU/index.asp].

[77] Standard Operating Procedures E, [http://www.fsis.usda.gov/Regulations_&_Policies/ Labeling_ FDA_MOU/index.asp].

[78] Standard Operating Procedures C, [http://www.fsis.usda.gov/Regulations_&_Policies/ Labeling_ FDA_MOU/index.asp].

[79] Safe and Suitable Ingredients Used in the Production of Meat and Poultry Products, FSIS Directive 7120.1, December 17, 2002, part V, [http://www.fsis.usda.gov/OPPDE/rdad/ FSISDirectives/7 120.1 .pdf].

[80] *Id.*

[81] Guidance on the Procedures for Joint Food Safety and Inspection Service and Food and Drug Administration Approval of Ingredients and Sources of Radiation Used in the Production of Meat and Poultry Products, [http://www.fsis.usda.gov/OPPDE/rdad/FRPubs/ 00-022N/Approvaloflngredients.htm].

[82] See Guidance on the Procedures for Joint FSIS and FDA Approval of Ingredients and Sources of Radiation Used in the Production of Meat and Poultry Products, [http://www.fsis.usda.gov/ OPPDE/rdad/FRPubs/00-022N/Approvaloflngredients.htm].

[83] *Id.*

[84] *Id.*

[85] *Id.*

[86] See 9 C.F.R. § 424.23.

[87] See 9 C.F.R. § 424.2 1(b)(3).

[88] FSIS has accepted data that has not been specifically applied to all categories if the data can be easily extrapolated to all species. *See* GRAS Notice No. 83, *supra* note 1.

[89] *Id.*

[90] Safe and Suitable Ingredients Used in the Production of Meat and Poultry Products, FSIS Directive 7120.1, December 17, 2002, part V, [http://www.fsis.usda.gov/OPPDE/rdad/ FSISDirectives/7 120.1 .pdf].

[91] *Id.*

[92] Specific substances that FSIS has approved by regulation are listed in 9 C.F.R. § 424.21(c).

[93] Safe and Suitable Ingredients Used in the Production of Meat and Poultry Products, FSIS Directive 7120.1, Dec. 17, 2002, [http://www.fsis.usda.gov/OPPDE/rdad/FSISDirectives/ 7120.1.pdf]. An amended version (dated Oct. 23, 2007) can be found at [http://www.fsis. usda.gov/OPPDE/rdad/FSISDirectives/7120.1Amend13.pdf].

[94] Guidance on the Procedures for Joint FSIS and FDA Approval of Ingredients and Sources of Radiation Used in the Production of Meat and Poultry Products, [http://www.fsis. usda.gov/OPPDE/rdad/FRPubs/00-022N/ApprovalofIngredients.htm].

[95] See 21 U.S.C. § 601 et seq.; 21 U.S.C. § 451 et seq.

[96] Guidance on the Procedures for Joint FSIS and FDA Approval of Ingredients and Sources of Radiation Used in the Production of Meat and Poultry Products, [http://www.fsis. usda.gov/OPPDE/rdad/FRPubs/00-022N/ApprovalofIngredients.htm].

[97] *Id.*

[98] The company first submitted a notification regarding tasteless smoke as a GRAS substance in 1998, but asked the FDA to cease to evaluate the notice and then resubmitted the notice in 1999. *See* GRAS Notice No. 15, *supra* note 3; Agency Response Letter GRAS Notice No. GRN 000005 from Linda S. Kahl, Regulatory Policy Branch, Division of Product Policy, Office of Premarket Approval, Center for Food Safety and Applied Nutrition, FDA, to Martin J. Hahn, Hogan & Hartson LLP (on behalf of Hawaii International Seafood, Inc.) (Dec. 11, 1998), [http://www.cfsan.fda.gov/~rdb/opa-g005. html].

[99] GRAS Notice No. 15, *supra* note 3.

[100] *See* GRAS Notice No. 83, *supra* note 1.

[101] *See* GRAS Notice No. 83, *supra* note 1.

[102] GRAS Notice No. 83, *supra* note 1.

[103] GRAS Notice No. 83, *supra* note 1. FDA response letters include FSIS conclusions on the substance. Although FSIS did not approve the use of carbon monoxide based on the original information provided, it did ultimately approve the use after receiving additional data. See Letters from Robert C. Post, Director, Labeling and Consumer Protection Staff, FSIS, to Lane Highbarger, Office of Food Additive Safety, FDA (Apr. 28, 2004 and June 2, 2004).

[104] *See* GRAS Notice No. 83, *supra* note 1.

[105] *See* Agency Response Letter GRAS Notice No. GRN 000143 from Laura M. Tarantino, Director, Office of Food Additive Safety, Center for Food Safety and Applied Nutrition, FDA, to Gary J. Kushner and Anne M. Boeckman, Hogan and Hartson (on behalf of Precept Foods, LLC) (July 29, 2004), [http://www.cfsan.fda.gov/~rdb/opa-g143.html], [hereinafter GRAS Notice No. 143]; Agency Response Letter GRAS Notice No. GRN 000167, from Laura M. Tarantino, Director, Office of Food Additive Safety, Center for Food Safety and Applied Nutrition, FDA, to Mark L. Itzkoff, Olsson, Frank and Weeda, PC (on behalf of Tyson Foods, Inc.), [http://www.cfsan.fda.gov/~rdb/opa-g1 67 .html], [hereinafter GRAS Notice No. 167]. The FDA ceased to evaluate another GRAS notification regarding carbon monoxide, GRAS Notice No. 167, at the request of the notifier, Freezing Machines, Inc. Agency Response Letter GRAS Notice No. GRN 000166 from Antonia Mattia, Director, Division of Biotech and GRAS Notice Review, Office of Food Additive Safety, Center for Food Safety and Applied Nutrition, FDA, to Mark L. Itzkoff, Olsson, Frank and Weeda, PC (on behalf of Freezing Machines, Inc.) (Oct. 12, 2005), [http://www.cfsan.fda.gov/~rdb/opag1 66.html].

[106] See listings for GRAS Notifications 188 and 194. Center for Food Safety and Applied Nutrition, FDA, Numerical Listing of GRAS Notices (Oct. 2007), [http://www.cfsan.fda. gov/~rdb/opa-gras.html].

[107] H.R. 3115, 110th Cong. § 3 (2007); H.R. 3610, 110th Cong. § 14 (2007).

[108] *See* H.R. 3115 § 3(a); H.R. 3610 § 14(a).

[109] The statement provided by the bills reads as follows: "SAFETY NOTICE: Carbon monoxide has been used to preserve the color of this product. Do not rely on color or the 'use or freeze by' date alone to judge the freshness or safety of the product. Discard any product with an unpleasant odor, slime, or a bulging package." H.R. 3115 § 3(a); H.R. 3610 § 14(a).

The statement provided by the discussion draft would remove the last sentence and slightly alter the wording as follows: "CONSUMER NOTICE: Carbon monoxide has been used to preserve the color of this product. Do not rely on color or the 'use or freeze by' date alone to judge the freshness of the product." Discussion Draft, § 132.

[110] 21 C.F.R. § 170.35(c)(2).

[111] 21 C.F.R. § 170.35(c)(4).

[112] 21 C.F.R. § 170.35(c)(5).

[113] 21 C.F.R. § 170.35(c)(4)-(6).

In: The FDA and Generally Recognized as Safe … ISBN: 978-1-61209-924-8
Editor: James C. Monahan, pp. 91 - 99

Chapter 3

GUIDANCE FOR INDUSTRY: FREQUENTLY ASKED QUESTIONS ABOUT GRAS*

James C. Monahan

This guidance represents the Food and Drug Administration's (FDA's) current thinking on this topic. It does not create or confer any rights for or on any person and does not operate to bind FDA or the public. An alternative approach can be used if such approach satisfies the requirements of the applicable statutes and regulations.

INTRODUCTION

This list of frequently asked questions (FAQ) is intended to be a convenient place to find answers to common questions about the food ingredient classification known as "generally recognized as safe" or "GRAS." This FAQ addresses common questions about the regulatory process and regulatory considerations regarding whether the use of a food substance is GRAS.

1. What does "GRAS" mean?
2. What are the criteria for GRAS status?

---

* This is an edited, reformatted and augmented edition of a United States Department of Health and Human Services, Food and Drug Administration, Center for Food Safety amd Applied Nutrition publication, dated December 2004.

3. In what way are the criteria for the use of a substance to be GRAS similar to that for the approved use of a food additive?
4. In what way are the criteria for the use of a substance to be GRAS different from that for the approved use of a food additive?
5. If an ingredient is GRAS for one use, is it GRAS for all uses?
6. Is a substance that is used to impart color eligible for classification as GRAS?
7. Is a substance that is used as a dietary ingredient of a dietary supplement eligible for classification eligible for classification as GRAS?
8. Must FDA approve GRAS substances?
9. What is GRAS affirmation?
10. Does FDA currently have a program to affirm that one or more uses of a food substance are GRAS?
11. What is the GRAS notification program?
12. If I choose to notify FDA of my GRAS determination, how do I do so?
13. Where do I send my GRAS notice?
14. If I submit a GRAS notice, how long will it take for me to receive a response from FDA?
15. If I submit a GRAS notice about a food substance, must I wait until I receive a response from FDA before I market that substance?
16. Does FDA have a list of substances that are used in food on the basis of the GRAS provision?
17. Can the use of a substance be GRAS even if it is not listed by FDA?

1. What does "GRAS" Mean?

"GRAS" is an acronym for the phrase Generally Recognized As Safe. Under sections 201(s) and 409 of the Federal Food, Drug, and Cosmetic Act (the Act), any substance that is intentionally added to food is a food additive, that is subject to premarket review and approval by FDA, unless the substance is generally recognized, among qualified experts, as having been adequately shown to be safe under the conditions of its intended use, or unless the use of the substance is otherwise excluded from the definition of a food additive. For example, substances whose use meets the definition of a pesticide, a dietary ingredient of a dietary supplement, a color additive, a new animal drug, or a substance approved for such use prior to September 6, 1958, are excluded from the definition of food additive. Sections 201(s) and 409 were enacted in 1958 as part of the Food Additives Amendment to the Act. While it is impracticable to list all ingredients whose use is generally recognized as safe, FDA published a partial list of food ingredients whose use is generally recognized as safe to aid the industry's understanding of what did not require approval.

2. What are the Criteria for GRAS Status?

Under sections 201(s) and 409 of the Act, and FDA's implementing regulations in 21 CFR 170.3 and 21 CFR 170.30, the use of a food substance may be GRAS either through scientific procedures or, for a substance used in food before 1958, through experience based on common use in food.

- Under 21 CFR 170.30(b), general recognition of safety through scientific procedures requires the same quantity and quality of scientific evidence as is required to obtain approval of the substance as a food additive and ordinarily is based upon published studies, which may be corroborated by unpublished studies and other data and information.
- Under 21 CFR 170.30(c) and 170.3(f), general recognition of safety through experience based on common use in foods requires a substantial history of consumption for food use by a significant number of consumers.

3. In what Way are the Criteria for the Use of a Substance to be GRAS Similar to that for the Approved Use of a Food Additive?

Regardless of whether the use of a substance is a food additive use or is GRAS, there must be evidence that the substance is safe under the conditions of its intended use. FDA has defined "safe" (21 CFR 170.3(i)) as a reasonable certainty in the minds of competent scientists that the substance is not harmful under its intended conditions of use. The specific data and information that demonstrate safety depend on the characteristics of the substance, the estimated dietary intake, and the population that will consume the substance.

4. In what Way are the Criteria for the Use of a Substance to be GRAS Different from that for the Approved Use of a Food Additive?

A GRAS substance is distinguished from a food additive on the basis of the common knowledge about the safety of the substance for its intended use. As FDA discussed in a proposed rule to establish a voluntary notification program for GRAS substances (62 Fed. Reg. 18938; April 17, 1997), the data and information relied on to establish the safety of the use of a GRAS substance must be generally available (e.g., through publication in the scientific literature) and there must be a basis to conclude that there is consensus among qualified experts about the safety of the substance for its intended use. Thus, the difference between use of a food additive and use of a GRAS substance relates to the widespread awareness of the data and information about the substance, i.e., who has access to the data and information and who has reviewed those data and information.

- For a food additive, privately held data and information about the use of the substance are sent by the sponsor to FDA and FDA evaluates those data and information to determine whether they establish that the substance is safe under the conditions of its intended use (21 CFR 171.1).
- For a GRAS substance, generally available data and information about the use of the substance are known and accepted widely by qualified experts, and there is a basis to conclude that there is consensus among qualified experts that those data and information establish that the substance is safe under the conditions of its intended use. (proposed 170.36 (c)(4)(i)(C))

5. If an Ingredient is GRAS for One Use, is it GRAS for All Uses?

Not necessarily. Under section 201(s) of the Act, it is the use of a substance, rather than the substance itself, that is eligible for the GRAS exemption (62 Fed. Reg. 18939; April 17, 1997). A determination of the safety of the use of an ingredient includes information about the characteristics of the substance, the estimated dietary intake under the intended conditions of use, and the population that will consume the substance (proposed 21 CFR 170.36 (c)(1)(iii)). Dietary intake of a substance depends on the food categories in which it will be used and the level of use in each of those food categories. For information about how FDA estimates dietary intake of a food substance, see FDA's document entitled

"Estimating Exposure to Direct Food Additives And Chemical Contaminants in the Diet" [August 2006: See updated information "Guidance for Industry - Estimating Dietary Intake of Substances in Food"]. Some uses of a food substance are intended for a narrowly defined population, such as newborn infants who consume infant formula as the sole item of the diet; in such a circumstance, there may be special considerations associated with that population but not with general use of the food substance.

6. Is a Substance that is Used to Impart Color Eligible for Classification as GRAS?

The short answer is "No." Under section 201(s) of the Act, the GRAS provision applies to the definition of a food additive. There is no corresponding provision in the definition (in section 201(t) of the Act) of a color additive.

However, under section 201(t)(1) and 21 CFR 70.3(f), the term color additive means a material that is a dye, pigment, or other substance made by a process of synthesis or similar artifice, or extracted, isolated, or otherwise derived from a vegetable, animal, mineral, or other source, and that is capable (alone or through reaction with another substance) of imparting color when added or applied to a food; except that such term does not include any material which FDA, by regulation, determines is used (or intended to be used) solely for a purpose or purposes other than coloring. Under 21 CFR 70.3(g), a material that otherwise meets the definition of color additive can be exempt from that definition on the basis that it is used or intended to be used solely for a purpose or purposes other than coloring, as long as the material is used in a way that any color imparted is clearly unimportant insofar as the appearance, value, marketability, or consumer acceptability is concerned. Given the construct of section 201(t)(1) of the Act and 21 CFR 70.3(f) and (g), the use of a substance that is capable of imparting color may constitute use as both a color additive and as a food additive or GRAS substance. For example, beta-carotene is both approved for use as a color additive (21 CFR 73.95) and affirmed as GRAS for use as a nutrient (21 CFR 184.1245); in some food products, beta-carotene may be used for both purposes.

7. Is a Substance that is Used as a Dietary Ingredient of a Dietary Supplement Eligible for Classification as GRAS?

Under section 201(s) of the Act, the ingredients whose use is GRAS are excluded from the definition of a food additive. That definition of food additive also specifies that the term "food additive" does not include a dietary ingredient of a dietary supplement described in section 201(ff) of the Act or intended for use in a dietary supplement. Thus, it is meaningless to refer to a GRAS exclusion from the food additive definition for dietary ingredients that are already excluded from that definition. However, some dietary ingredients that may be used in a dietary supplement may also be GRAS for use in a conventional food (e.g., vitamin C; calcium carbonate).

8. Must FDA Approve GRAS Substances?

No. If the use of a food substance is GRAS, it is not subject to the premarket review and approval requirement by FDA.

9. What is GRAS Affirmation?

GRAS affirmation is a process that FDA developed in the 1970s. In response to concerns raised by new information on cyclamate salts, then-President Nixon directed FDA to re-examine the safety of substances considered to be GRAS. FDA announced that the agency would evaluate, by contemporary standards of the time, the available safety information regarding substances considered to be GRAS. If the revaluation of current data confirmed that use was GRAS, FDA would promulgate a new GRAS regulation, affirming that finding. FDA also established procedures whereby an individual could petition FDA to review the GRAS status of substances that would not have been considered as part of the agency's GRAS review.

10. Does FDA Currently Have a Program to Affirm that One or More Uses of a Food Substance are GRAS?

In a proposed rule that FDA published in 1997 (62 Fed. Reg. 18938; April 17, 1997), FDA explained why the agency could no longer devote resources to the voluntary GRAS affirmation petition process that is described in 21 CFR 170.35(c) and proposed to abolish that process and replace it with a notification procedure. The agency has not yet issued a final rule however, and the petition procedure remains in the agency's regulations. However, at this time FDA is not committing resources to the review of GRAS affirmation petitions.

11. What is the GRAS Notification Program?

The GRAS notification program is a voluntary procedure that is operating under a proposed rule issued in 1997 (62 Fed. Reg. 18938; April 17, 1997). The notification program is intended to replace the GRAS affirmation process by providing a mechanism whereby a person may inform FDA of a determination that the use of a substance is GRAS, rather than petition FDA to affirm that the use of a substance is GRAS. The submitted notice includes a "GRAS exemption claim" that includes a succinct description of the substance, the applicable conditions of use, and the statutory basis for the GRAS determination (i.e., through scientific procedures or through experience based on common use in food). A GRAS notice also includes information about the identity and properties of the notified substance and a discussion of the notifier's reasons for concluding that the substance is GRAS for its intended use.

12. If I Choose to Notify FDA of My GRAS Determination, How do I do So?

FDA described the procedure for submitting a GRAS notice in the proposed rule to establish the notification procedure (62 Fed. Reg. 18938; April 17, 1997). Because the proposed rule is a lengthy document, our Internet site has a specific link to the part of the proposed rule that describes the procedure. You can find both the complete proposed rule and the link to the procedure on the main page of the GRAS notification program.

13. Where do I Send My GRAS Notice?

You should send your GRAS notice to the Office of Food Additive Safety (HFS-255), Center for Food Safety and Applied Nutrition, Food and Drug Administration, 5100 Paint Branch Parkway, College Park, MD 20740. [Note that our office moved since we issued the proposed rule to establish a GRAS notification procedure and, thus, the address where you should send your GRAS notice is different from the address that we published in the proposed rule that describes the procedure].

14. If I Submit a GRAS Notice, How Long Will it Take for Me to Receive a Response from FDA?

Our goal is to respond to most GRAS notices within 180 days.

15. If I Submit a GRAS Notice About a Food Substance, Must I Wait Until I Receive a Response from FDA before I Market that Substance?

No. If one is correct in determining that the intended use of an ingredient is GRAS, use of the ingredient is not subject to any legal requirement for FDA review and approval. Your decision to submit a GRAS notice is voluntary, and FDA's response to a GRAS notice is not an approval. You may market a substance that you determine to be GRAS for a particular use without informing FDA or, if FDA is so informed, while FDA is reviewing that information (62 Fed. Reg. 18951; April 17, 1997). We recognize, however, that some firms prefer to know that FDA has reviewed its notice of a GRAS determination, without raising safety or legal issues, before marketing.

16. Does FDA have a List of Substances that are Used in Food on the Basis of the GRAS Provision?

FDA has several lists of GRAS substances. Importantly, these lists are not all-inclusive. Because the use of a GRAS substance is not subject to premarket review and approval by FDA, it is impracticable to list all substances that are used in food on the basis of the GRAS provision.

- 21 CFR Part 182 contains the remnants of a list, which FDA established in its regulations shortly after passage of the 1958 Food Additives Amendment. The list is organized according to the intended use of these substances. As part of the agency's comprehensive review of GRAS substances in the 1970s, FDA affirmed that the use of some of the ingredients on this original GRAS list is GRAS, and moved the affirmed uses of the substance to 21 CFR Part 184.
- 21 CFR Part 184 contains a list of substances that FDA affirmed as GRAS as direct food ingredients for general or specific uses. This list derives from the agency's 1970s comprehensive review of GRAS

substances and from petitions that FDA received to affirm the GRAS status of particular uses of some food ingredients.

- 21 CFR Part 186 contains a list of substances that FDA affirmed as GRAS for certain indirect food uses.
- FDA's Internet site also contains a list of substances that have been the subject of a notice to FDA - i.e., when a firm has notified FDA about its view that a particular use of a substance is GRAS. You can access this summary of GRAS notices, along with FDA's response, from the GRAS Notification Program page.

17. Can the Use of a Substance be GRAS Even if it is not Listed by FDA?

Yes. Because the use of a GRAS substance is not subject to premarket review and approval by FDA, it is impracticable to list all substances that are used in food on the basis of the GRAS provision (21 CFR 182.1). The use of a substance is GRAS because of widespread knowledge among the community of qualified experts, not because of a listing or other administrative activity.

Note: The above list of Frequently Asked Questions was taken from the following FDA website:

http://www.fda.gov/Food/GuidanceComplianceRegulatoryInformation/GuidanceDocuments/FoodIngredientsandPackaging/ucm061846.html

Chapter Sources

Chapter 1 - This is an edited, reformatted and augmented version of a U.S. Government Accountability Office publication, GAO-10-246, dated February, 2010.

Chapter 2 - This is an edited, reformatted and augmented version of a Congressional Research Service publication, RL34247, dated September 25, 2008.

Chapter 3 - This is an edited, reformatted and augmented edition of a United States Department of Health and Human Services, Food and Drug Administration, Center for Food Safety amd Applied Nutrition publication, dated December 2004.

INDEX